# THE PRACTICALITY OF PULSED FAST NEUTRON TRANSMISSION SPECTROSCOPY FOR AVIATION SECURITY

**Panel on Assessment of the Practicality of Pulsed Fast Neutron Transmission Spectroscopy for Aviation Security**

National Materials Advisory Board
Commission on Engineering and Technical Systems
National Research Council

NMAB-482-6
Washington, D.C. 1999

**NATIONAL ACADEMY PRESS • 2101 Constitution Avenue, N.W. • Washington, D.C. 20418**

**NOTICE**: The project that is the subject of this report was approved by the Governing Board of the National Research Council, whose members are drawn from the councils of the National Academy of Sciences, the National Academy of Engineering, and the Institute of Medicine. The members of the committee responsible for the report were chosen for their special competencies and with regard for appropriate balance.

This report has been reviewed by a group other than the authors according to procedures approved by a Report Review Committee consisting of members of the National Academy of Sciences, the National Academy of Engineering, and the Institute of Medicine.

The National Academy of Sciences is a private, nonprofit, self-perpetuating society of distinguished scholars engaged in scientific and engineering research, dedicated to the furtherance of science and technology and to their use for the general welfare. Upon the authority of the charter granted to it by the Congress in 1863, the Academy has a mandate that requires it to advise the federal government on scientific and technical matters. Dr. Bruce Alberts is president of the National Academy of Sciences.

The National Academy of Engineering was established in 1964, under the charter of the National Academy of Sciences, as a parallel organization of outstanding engineers. It is autonomous in its administration and in the selection of its members, sharing with the National Academy of Sciences the responsibility for advising the federal government. The National Academy of Engineering also sponsors engineering programs aimed at meeting national needs, encourages education and research, and recognizes the superior achievements of engineers. Dr. William Wulf is president of the National Academy of Engineering.

The Institute of Medicine was established in 1970 by the National Academy of Sciences to secure the services of eminent members of appropriate professions in the examination of policy matters pertaining to the health of the public. The Institute acts under the responsibility given to the National Academy of Sciences by its congressional charter to be an adviser to the federal government and, upon its own initiative, to identify issues of medical care, research, and education. Dr. Kenneth I. Shine is president of the Institute of Medicine.

The National Research Council was organized by the National Academy of Sciences in 1916 to associate the broad community of science and technology with the Academy's purposes of furthering knowledge and advising the federal government. Functioning in accordance with general policies determined by the Academy, the Council has become the principal operating agency of both the National Academy of Sciences and the National Academy of Engineering in providing services to the government, the public, and the scientific and engineering communities. The Council is administered jointly by both Academies and the Institute of Medicine. Dr. Bruce Alberts and Dr. William Wulf are chairman and vice chairman, respectively, of the National Research Council.

This study by the National Materials Advisory Board was conducted under Contract No. DTFA03-94-C-00068 with the Federal Aviation Administration. Any opinions, findings, conclusions, or recommendations expressed in this publication are those of the author(s) and do not necessarily reflect the views of the organizations or agencies that provided support for the project.

Available in limited supply from:
National Materials Advisory Board
2101 Constitution Avenue, N.W.
HA-262
Washington, DC 20418
202-334-3505
nmab@nas.edu

Additional copies are available for sale from:
National Academy Press
Box 285
2101 Constitution Ave., N.W.
Washington, DC 20055
800-624-6242
202-334-3313 (in the Washington Metropolitan Area)
http://www.nap.edu

International Standard Book Number: 0-309-06449-X

Printed in the United States of America.

## PANEL ON ASSESSMENT OF THE PRACTICALITY OF PULSED FAST NEUTRON TRANSMISSION SPECTROSCOPY FOR AVIATION SECURITY

PATRICK J. GRIFFIN (chair), Sandia National Laboratories, Albuquerque, New Mexico
ROBERT BERKEBILE, consultant, Leesburg, Florida
HOMER BOYNTON, consultant, Hilton Head Island, South Carolina
LEN LIMMER, consultant, Fort Worth, Texas
HARRY MARTZ, Lawrence Livermore National Laboratory, Livermore, California
CLINTON OSTER, JR., Indiana University, Bloomington

### National Materials Advisory Board Liaison

JAMES WAGNER, Case Western Reserve University, Cleveland, Ohio

### National Materials Advisory Board Staff

RICHARD CHAIT, director
CHARLES T. HACH, staff officer
SANDRA HYLAND, senior program manager (until June 1998)
JANICE M. PRISCO, project assistant

### Government Liaisons

JOHN DALY, U.S. Department of Transportation, Washington, D.C.
ANTHONY FAINBERG, Federal Aviation Administration, Washington, D.C.
PAUL JANKOWSKI, Federal Aviation Administration Technical Center, Atlantic City, New Jersey
LYLE MALOTKY, Federal Aviation Administration, Washington, D.C.
ALAN K. NOVAKOFF, Federal Aviation Administration Technical Center, Atlantic City, New Jersey

## NATIONAL MATERIALS ADVISORY BOARD

# Preface

The Federal Aviation Administration (FAA) of the U.S. Department of Transportation was established in 1958 to promote and ensure the safety of air travel. One objective of the FAA is to reduce the vulnerability of the civil air transport system to terrorist threats by employing procedural and technical means to detect and counter threats. The role of the FAA in aviation security also includes developing new technologies for aviation security through the FAA's research and development program.

One area of research being pursued by the FAA is accelerator-based nuclear technologies that detect explosives by measuring the elemental composition of the material under examination. Pulsed fast neutron transmission spectroscopy (PFNTS) is one of these element-specific detection technologies. PFNTS, however, has a number of practical limitations, including large size and weight, the necessity of radiation shielding, and the regulatory and safety issues associated with using neutron-producing equipment in an airport environment.

In the second interim report of the National Research Council's (NRC) Committee on Commercial Aviation Security (CCAS), the committee recommended that the FAA not pursue accelerator-based technologies for primary screening of checked baggage and not fund development projects for large accelerator-based hardware. The CCAS concluded that the detection performance of these methods should be better understood before the FAA addressed airport integration issues. In 1994, the FAA awarded Tensor Technology a two-year grant to build a multidimensional neutron radiometer (MDNR) airline security system. The detection performance of the MDNR showed that it could potentially meet the probability of detection ($P_d$) required for FAA certification for all but one of the required explosives categories. Based on these test results and in light of the recommendations of the CCAS, the FAA awarded Tensor a six-month cooperative agreement grant to present the company's evaluation of PFNTS compared to other, currently available technologies for the primary screening of passenger baggage for explosives and for the screening of cargo in airports.

In 1998, the FAA requested that the NRC review and evaluate Tensor Technology's assessment of PFNTS in light of the CCAS's recommendations and technical developments since the second interim report. In response to the FAA's request, the NRC convened the Panel on Assessment of the Practicality of Pulsed Fast Neutron Transmission Spectroscopy for Aviation Security under the auspices of the CCAS. The panel was charged with evaluating the practicality of PFNTS for primary screening of passenger baggage or for screening cargo, as compared to currently available x-ray computed tomography (CT)-based systems.

This report evaluates the practicality of PFNTS for aviation security under current performance requirements, as compared to FAA-certified x-ray CT-based systems. The panel also provides several recommendations for prioritizing research to address the technical limitations of PFNTS in the event that funds are appropriated for the continued development of this technology. It should be noted that the panel does not support or oppose such appropriations. It should also be noted that solving the technical challenges of PFNTS will not address the practical limitations (e.g., size and weight) of this technology, which may be the most important factors in determining the role of PFNTS in aviation security.

Patrick J. Griffin, chair
Panel on Assessment of the Practicality
of Pulsed Fast Neutron Transmission
Spectroscopy for Aviation Security

# Acknowledgments

The Panel on Assessment of the Practicality of Pulsed Fast Neutron Transmission Spectroscopy for Aviation Security would like to acknowledge the individuals who contributed to this study, including the following speakers: Curtis Bell, Federal Aviation Administration; Anthony Fainberg, Federal Aviation Administration; Richard Lanza, Massachussetts Institute of Technology; Thomas "Gill" Miller, Tensor Technology; John Overley, University of Oregon; Fred Roder, Federal Aviation Administration; and Peter K. Van Staagan, Tensor Technology. The panel is also grateful for the contributions of the contracting office technical representatives, Paul Jankowski and Alan K. Novakoff. In addition, the panel is appreciative of the insights provided by John Daly, U.S. Department of Transportation; Rodger Dickey, Dallas/Fort Worth International Airport; Karl Erdman, Ebco Technology; Ronald Krauss, Federal Aviation Administration; Lyle Malotky, Federal Aviation Administration; Ronald Polillo, Federal Aviation Administration; and Johannes E. van Lier, University De Sherbrooke.

This report has been reviewed in draft form by individuals chosen for their diverse perspectives and technical expertise, in accordance with procedures approved by the NRC's Report Review Committee. The purpose of this independent review is to provide candid and critical comments that will assist the institution in making the published report as sound as possible and to ensure that the report meets institutional standards for objectivity, evidence, and responsiveness to the study charge. The review comments and draft manuscript remain confidential to protect the integrity of the deliberative process. We wish to thank the following individuals for their participation in the review of this report: Jack Bullard, American Airlines; Robert Gagne, Food and Drug Administration; Robert E. Green, Johns Hopkins University; James Hall, Lawrence Livermore National Laboratory; John LaRue, Dallas/Fort Worth International Airport; Hyla Napadensky, Napadensky Energetics (retired); and John Strong, College of William and Mary. While the individuals listed above have provided constructive comments and suggestions, it must be emphasized that responsibility for the final content of this report rests entirely with the authoring committee and the NRC.

For organizing panel meetings and directing this report to completion, the panel would like to thank Charles Hach, Sandra Hyland, Lois Lobo, Janice Prisco, and Pat Williams, staff members of the National Materials Advisory Board. The panel is also appreciative of the efforts of Carol R. Arenberg, editor, Commission on Engineering and Technical Systems.

# Acronyms

| | |
|---|---|
| CCAS | Committee on Commercial Aviation Security |
| CFR | Code of Federal Regulations |
| CT | computed tomography |
| EDS | explosives-detection system |
| EIS | Environmental Impact Statement |
| FAA | Federal Aviation Administration |
| MDNR | multidimensional neutron radiometer |
| NMAB | National Materials Advisory Board |
| NRC | National Research Council |
| $P_d$ | probability of detection |
| $P_{fa}$ | probability of false alarm |
| PFNTS | pulsed fast neutron transmission spectroscopy |
| RCRA | Resource Conservation and Recovery Act |
| SEIPT | Security Equipment Integrated Product Team |
| TLD | thermoluminescent dosimeter |

# Contents

# Tables, Figures, and Boxes

## TABLES



## FIGURES



## BOXES

# Executive Summary

The White House Commission on Aviation Safety and Security recommended the deployment of explosives-detection equipment, including x-ray computed tomography (CT) based explosives-detection systems (EDSs), the only detection method that has passed certification testing by the Federal Aviation Administration (FAA). Several other technologies are under development and have been tested in laboratory settings but have not passed certification testing.

One area of research being pursued by the FAA is accelerator-based nuclear detection technologies that detect explosives by measuring the elemental composition of the material under examination. These technologies exploit the high nitrogen and oxygen content present in most explosives. Pulsed fast neutron transmission spectroscopy (PFNTS), one of these element-specific detection technologies, also has the potential for generating low-resolution tomographic images (NRC, 1998; Overley, 1987). However, PFNTS also has a number of practical limitations, including large size and weight, the necessity of radiation shielding, and the regulatory and safety issues associated with using a nuclear-based technology (NRC, 1993, 1997).

## BACKGROUND

In 1993, the FAA requested that the National Research Council (NRC) assist the agency in assessing its explosives-detection program. The NRC responded to this request by convening the Committee on Commercial Aviation Security (CCAS), which has produced two interim reports (NRC, 1996, 1997) containing recommendations for structuring the research portfolio for the FAA's explosives-detection program. The committee's recommendations addressed bulk explosives detection, trace explosives detection, combined technologies, and human factors. In the second interim report (NRC, 1997), the CCAS recommended that the FAA not pursue accelerator-based technologies for primary screening of checked baggage and not fund development projects for large accelerator-based hardware. The CCAS concluded that the detection performance of these methods should be better understood before the FAA addressed airport integration issues and recommended that existing laboratory equipment be used to determine the detection limits of PFNTS for Class A[1] explosives (NRC, 1997).

In 1994, the FAA awarded Tensor Technology a two-year grant to build a multidimensional neutron radiometer (MDNR) airline security system. This project included building an airline security system, transporting the system to a nuclear accelerator, and testing the MDNR to determine its sensitivity for detecting explosives concealed in suitcases (Tensor Technology, 1998). The detection performance of the MDNR showed that it could potentially meet the probability of detection required for FAA certification for all but one of the required explosives categories. Based on these test results and in light of the recommendations of the CCAS, the FAA awarded Tensor a six-month cooperative agreement grant to compare the performance of PFNTS with the performance of other, currently available technologies for primary screening of passenger baggage for explosives and for screening cargo in airports.

In 1998, the FAA requested that the NRC review and evaluate Tensor Technology's assessment of PFNTS in light of the CCAS's recommendations (see Box ES-1) and technical developments since the second interim report. In response to the FAA's request, the NRC convened the Panel on Assessment of the Practicality of Pulsed Fast Neutron Transmission Spectroscopy for Aviation Security under the auspices of the CCAS. The panel was charged with evaluating

[1] Class A and Class B explosives are categories devised by the panel and do not represent a designation made by the FAA. PFNTS has difficulty detecting certain types and configurations of explosives defined as Class A explosives in this report. Class B explosives include all other explosives in the FAA's certification test set. A detailed description of Class A and Class B explosives is not available in this report due to the sensitive nature of this information. Specific questions regarding the performance of PFNTS should be addressed to the FAA.

**BOX ES-1**
**CCAS Recommendations for Accelerator-Based Explosives-Detection Technologies**

**Recommendation 4-1. Do not consider accelerator-based technologies to have promise for deployment as a primary screening procedure for checked baggage inspection.** Any screening procedure relying on an accelerator cannot compete with available technologies on either cost or practicality bases.

**Recommendation 4-2. Do not fund any large accelerator-based hardware development projects.** Combinations of experimental work with existing laboratory equipment, mathematical modeling, and simulation can better define the potential of the nuclear technologies without the expense or time required to design and build new hardware.

Source: NRC, 1997.

the practicality of PFNTS for primary screening of passenger baggage or for screening cargo, compared to currently available x-ray CT-based systems.

## FINDINGS

The panel examined the principles of operation of PFNTS and the results of laboratory-based blind tests on explosives in cluttered passenger bags. Some PFNTS tests demonstrated detection levels consistent with the FAA's EDS certification standards, but two important deficiencies were revealed. First, PFNTS did not demonstrate an ability to detect Class A explosives, an important class of explosives that most alternative technologies also have problems detecting. Second, PFNTS, when used with a two-dimensional area neutron detector,[2] had a higher false alarm rate than the FAA's EDS certification criteria allow.

Tensor provided a conceptual design of a PFNTS-based explosives-detection device (the MDNR) for implementation in an airport rather than a laboratory setting. The panel found that the MDNR provided a reasonable baseline conceptual design for assessing a PFNTS-based explosives-detection device for airport implementation. One of the unique characteristics of the MDNR design is the use of a cyclotron rather than a linear accelerator, which was used in all previous PFNTS testing. The use of a cyclotron would reduce the size of PFNTS and, therefore, make it more practical for airport integration. However, the reduction in size entails a substantial increase in system weight attributable to the heavy magnets in the cyclotron.

One could argue that the accelerator weight of 20 tonnes (22 tons) is not a severe penalty compared to the 109-tonne (120-ton) weight of the cyclotron shield enclosure and the 528-tonne (581-ton) weight of the vault enclosure required for any neutron-producing accelerator system. The panel identified size, weight, radiation shielding requirements, and complicated baggage flow constraints as problems that could arise during airport integration. The panel identified the use of commercial parts for the cyclotron and the detector electronics as an advantage of the MDNR over many other laboratory-based nuclear detection technologies.

The panel compared the practicality of the MDNR conceptual design to x-ray CT-based EDSs for implementation in airports. The laboratory-demonstrated explosives-detection performance of the MDNR was inferior to the laboratory-based performance of certified CT-based systems, both in the detection of Class A explosives and in the false alarm rate. The MDNR testing protocol differed from the CT system certification testing protocol, however, in the distribution of the subclasses of Class A explosives and the way false alarm statistics were collected. A direct comparison of the detection performance of x-ray CT and MDNR would require a significant increase in the statistical database for MDNR and a common test protocol. However, the available test results suggest that, even in the area of Class B explosives, the MDNR did not show a significant performance advantage over x-ray CT-based EDSs.

The resolution of false alarms is not part of the current EDS certification requirements, but it is an important consideration in the selection and fielding of equipment in airports. Laboratory test results indicate that the false alarm rate for PFNTS is between 13 and 25 percent. Evaluation of these results suggests that even an optimized PFNTS system would have a false alarm rate of at least 4 percent. In the absence of an acceptable alarm resolution protocol, this is an unacceptably high rate for airport implementation. Current x-ray CT systems in airports rely on operators to interpret high-resolution images to resolve automated alarms. Al-

[2] A two-dimensional area detector would be required for efficient bag throughput in airport operation.

though operator intervention could potentially lower the probability of detection for x-ray CT-based systems, it has been demonstrated to reduce the false alarm rate. The image produced by transmitted neutrons from a PFNTS-based device would not be sufficient for an operator to resolve automated false alarms. Unless a highly reliable alarm resolution method can be found, PFNTS would have to be combined with a high-resolution x-ray-based imaging technology for the purpose of alarm resolution.

The potential of PFNTS for screening small loose cargo packages has not been sufficiently explored using existing research accelerators. In addition, the characterization of air cargo (type, size, weight, delivery constraints, method of delivery to the airport) is not sufficient to develop a valid testing protocol, although the FAA provided the panel with a working threat definition (explosive type and quantity) to use in evaluating cargo inspection technologies. Based on existing cargo characterization data (FAA, 1996) and analytic estimates of neutron attenuation, the panel believes that PFNTS does not have a realistic potential for screening the full spectrum of cargo containers or pallets to this threat level. Cargo containers filled with cargo with high hydrogen content would attenuate the neutron transmission to a level comparable to the room neutron scattering background, thus rendering neutron transmission-based detection approaches ineffective. A thorough analysis of the potential of PFNTS for cargo scanning, however, will require a statistically significant set of explosives-detection test data with various types of cargo.

## CONCLUSIONS AND RECOMMENDATIONS

The greatest performance shortfall of PFNTS is its failure to detect Class A explosives. Unless this issue is resolved, PFNTS has no future as an explosives-detection technology in commercial aviation security. Even if this issue were resolved and the performance were equal to available FAA-certified EDSs, PFNTS in general—and the MDNR design specifically—has other disadvantages related to its size and weight that would preclude its selection for airport baggage scanning. Based on the FAA's current certification testing requirements for the detection of Class B explosives (as opposed to Class A explosives), PFNTS-based technologies would not be selected for primary screening of carry-on baggage, checked baggage, or cargo because of the difficulty of integrating these technologies into existing airport terminals and because of the safety issues associated with the operation of radiation-producing accelerators.

Tests have indicated that PFNTS has the potential for very low false alarm rates. If this potential were realized, then the PFNTS might play a role in aviation security but not as a primary EDS under current certification requirements. The panel concluded that only if the low false alarm rate is validated should the PFNTS-based system be taken through prototype development and demonstration in an airport environment. Even after airport testing, however, the system design would probably not be widely deployed in airports but would be placed on the shelf as a validated and characterized device that could be reevaluated as explosives threats changed or as regulatory requirements were refined.

### Questions Posed by the FAA

At the first meeting of the Panel on Assessment of the Practicality of PFNTS for Aviation Security, the FAA asked that the panel address four questions during the course of the study. The panel and the NRC staff determined that these questions fall within the panel's Statement of Task. The questions are addressed below.

**Question 1.** Given the choice, would airlines select the PFNTS instead of equipment based on currently available x-ray CT for checked baggage inspection?

**Answer.** No, the airlines would not choose a PFNTS-based explosives-detection device for three reasons. First, PFNTS has not demonstrated an ability to meet the FAA's certification requirements for detecting Class A explosives. Second, the area detector configuration has not demonstrated an ability to meet the FAA's false alarm requirements. Third, because of the difficulties in deployment and integration of a PFNTS-based device, including size, weight, and safety issues, the airlines would choose currently available x-ray CT-based EDSs.

**Question 2.** If so, is their preference for this technology strong enough to justify the remaining costs to develop this technology (estimated to be \$20 million to \$30 million)?

**Answer.** Not applicable.

**Question 3.** Does PFNTS have any realistic potential for application to full cargo container inspections?

**Answer.** No. Based on Tensor's analysis of PFNTS for cargo screening of LD-3 containers containing single items (a reasonable projection given the sparseness of existing test data), PFNTS could only interrogate 57 percent of air cargo shipped in LD-3 containers. The other 43 percent contains large amounts of hydrogenous and, therefore, highly neutron-attenuating material, rendering PFNTS ineffective for screening. Further complicating the use of PFNTS for cargo screening is that containerized cargo is not always uniform in composition. The likelihood of false alarms from nonhomogeneous containerized cargo assembled by cargo consolidators has not been determined. Finally, the capability of PFNTS to detect explosives concealed in thick containers (e.g., the LD-3) has not been experimentally verified. Current estimates on neutron attenuation (based on crude exponential algorithms) are sufficiently accurate to raise concerns about highly attenuating hydrogenous cargo. These estimates do not treat beam divergence from neutron scattering in the cargo and are, therefore, not sufficient to validate the PFNTS detection for low-attenuating scenarios.

**Question 4.** What experiments, if any, should be pursued in the near future to further define this potential?

**Answer.** Because PFNTS does not appear to be either practical or currently desirable for airport deployment, the panel does not recommend that experiments addressing the airport integration of PFNTS be pursued at this time. However, experimental verification of Tensor's simulation of PFNTS performance for cargo screening might be useful.

## Prototype

The inability of PFNTS-based explosives-detection technologies to detect Class A explosives at the probability of detection level required for EDS certification is a critical limitation. PFNTS-based techniques also demonstrated unacceptable false alarm rates when using area detectors and did not demonstrate a viable approach for resolving alarms. Unless and until these limitations are overcome, there is no reason for the FAA to pursue other technical or operational issues associated with integrating the technology into an airport setting.

**Recommendation.** The FAA should not fund the development of a prototype multidimensional neutron radiometer-based explosives-detection device.

**Recommendation.** At current levels of explosive threat and with the current state of the art, the FAA should not deploy pulsed fast neutron transmission spectroscopy-based explosives-detection technologies or devices for primary screening of carry-on baggage, checked baggage, or cargo.

**Recommendation.** At this stage, the FAA should not fund the development of an airport test facility.

## Research Priorities

Laboratory testing has not demonstrated the PFNTS technology to be technically desirable. However, because the threat to aviation security is dynamic, the requirements for explosives-detection systems certification may change over time. At some point, if existing deployed technologies do not provide adequate protection, a need for new explosives-detection approaches could arise. Among the technologies currently in development, PFNTS shows the most promise of meeting more stringent certification testing requirements because it is an element-specific detection technology. Therefore, even though the deployment of a PFNTS prototype designed for integration into an airport (e.g., MDNR) is not desirable at this time, valuable research could be conducted on the application of PFNTS technologies to explosives detection.

The research recommendations in this report are directed toward improving the detection performance of PFNTS-based explosives-detection technologies. However, even if these recommendations are followed and the detection performance is improved, practical limitations to the deployment of PFNTS-based devices remain (e.g., size, mass). These practical limitations should be taken into account when evaluating the potential of PFNTS for explosives detection in airports.

If funding becomes available for the development of PFNTS technology for explosives detection, the greatest benefit would be derived by addressing the current shortfalls of the technology rather than by developing and assembling a prototype MDNR unit for integration into an airport. Because the panel has not studied, and is not acquainted with, the whole spectrum of research requests for explosives-detection technologies submitted to the FAA, the panel neither supports nor opposes the allocation of research funds for further research on PFNTS.

**Recommendation.** The research priorities for pulsed fast neutron transmission spectroscopy should be directly related to the current shortfalls of the technology. Three major problem areas that should be addressed are listed below in order of importance:

1. detection of Class A explosives
2. reduction of false alarm rates
3. development of alarm resolution procedures

## Research Facilities

Consistent with the recommendations in the second interim report of the CCAS, this panel believes that if PFNTS continues to be pursued it should be researched and developed with current neutron sources, detector technology, and radiation modeling capabilities. However, it appears that no existing research facility can meet the requirements for demonstrating the potential of PFNTS for explosives detection. If efforts are made to acquire a more compatible research facility, the FAA should acquire an existing commercially available accelerator (rather than funding the development of a new accelerator).

If funding is allocated for the construction of a facility to develop PFNTS for explosives detection, the FAA would derive the greatest benefit from a facility that promotes broad cooperation and long-term multidisciplinary research. It should be noted that the panel does not support or oppose the allocation of funds for such a facility.

**Recommendation.** If the FAA acquires an accelerator to meet the PFNTS testing requirements, it should be configured to support a broad range of research activities.

## REFERENCES

FAA (Federal Aviation Administration). 1996. The Composition and Handling of Passenger Aircraft Cargo. DOT/FAA/AR-96/83. Washington, D.C.: Federal Aviation Administration.

NRC (National Research Council). 1998. Configuration Management and Performance Verification of Explosives-Detection Systems. National Materials Advisory Board. Washington, D.C.: National Academy Press. Available on the Internet: http://www.nap.edu/readingroom/enter2.cgi?0309061962.html

NRC. 1997. Second Interim Report of the Committee on Commercial Aviation Security. National Materials Advisory Board. Washington, D.C.: National Academy Press. Available on the Internet: http://www2.nas.edu/nmab/262e.html

NRC. 1996. First Interim Report of the Committee on Commercial Aviation Security. National Materials Advisory Board. Washington, D.C.: National Academy Press.

NRC. 1993. Detection of Explosives for Commercial Aviation Security. National Materials Advisory Board. Washington, D.C.: National Academy Press.

Overley, J.C. 1987. Element-sensitive computed tomography with fast neutrons. Nuclear Instruments and Methods in Physics Research B24/25: 1058–1062.

Tensor Technology, Inc. 1998. Advanced Studies on the Multi-Dimensional Neutron Radiometer (MDNR) Airline Security System. Washington, D.C.: Federal Aviation Administration.

# 1

# Introduction

Since the release of the widely publicized reports of the White House Commission on Aviation Safety and Security (1996, 1997)—which recommended the deployment of "significant numbers of computed tomography detection systems, upgraded x-rays, and other innovative systems"—the Federal Aviation Administration (FAA) has deployed 63[1] FAA-certified explosives-detection systems (EDSs),[2–4] four noncertified bulk-explosives detection devices, and more than 300 trace explosives-detection devices. To date, all of the FAA-certified EDSs employ an x-ray-based computed tomography (CT) detection methodology. However, through its Explosives Detection Program, the FAA continues to pursue research in other promising technologies for addressing the FAA's future needs for explosives detection.

## OVERVIEW OF PULSED FAST NEUTRON TRANSMISSION SPECTROSCOPY

One area of research the FAA has pursued is accelerator-based nuclear detection technologies that detect explosives by measuring the elemental composition of materials. These technologies exploit the high nitrogen and oxygen content found in most explosives. Pulsed fast neutron transmission spectroscopy (PFNTS) identifies explosives by the specific material- and energy-dependent absorption and scattering cross sections of neutrons interacting with the nuclei of different elements. PFNTS can determine the hydrogen, carbon, nitrogen, and oxygen content in an object, and the relative amounts of these elements can be used to discriminate explosive from nonexplosive materials. PFNTS also has the potential to generate low-resolution tomographic images (NRC, 1998; Overley, 1987). PFNTS also has a number of practical limitations, including large size and weight, the need for radiation shielding, and regulatory and safety issues associated with nuclear-based technologies (NRC 1993, 1997).

## BACKGROUND OF THIS STUDY

In 1993, the FAA requested that the National Research Council (NRC) assist the agency in assessing its explosives-detection program. The NRC responded to this request by convening the Committee on Commercial Aviation Security (CCAS). Since 1993, the committee has produced two interim reports (NRC, 1996, 1997) that provided recommendations for structuring the FAA's research portfolio for the explosives-detection program, including bulk explosives detection, explosives trace detection, combined technologies, and human factors. In the second interim report (NRC, 1997), the CCAS recommended that the FAA should not pursue accelerator-based technologies for the primary screening of checked baggage or fund the development of any large accelerator-based hardware (see Box 1-1).

The CCAS concluded that the detection performance of an explosives-detection method should be well understood before airport integration issues are addressed (NRC, 1997). Recent testing had indicated that, with the exception of Class A[5] explosives, the detection performance of PFNTS was con-

[1] The FAA intends to deploy 74 certified explosives-detection systems by March 1999.

[2] The following terminology is used throughout this report. An *explosives-detection device* is an instrument (not FAA certified) that incorporates a single detection method to detect one or more category of explosives. An *explosives-detection system (EDS)* is a self-contained unit composed of one or more integrated devices that has passed the FAA's explosive-detection certification test. *Explosives-detection equipment* is any equipment, certified or not, that can be used to detect explosives.

[3] Explosives-detection equipment includes any explosives-detection device or system that remotely senses some physical or chemical property of an object under investigation to determine if it is an explosive. Trace explosives-detection equipment requires that particles or vapor from the object under investigation be collected and identified.

[4] In this report, *explosives* include all forms and configurations of an explosive at threat level.

[5]Class A and Class B explosives are categories devised by the panel and do not represent a designation made by the FAA. PFNTS has difficulty detecting certain types and configurations of explosives defined as Class A

**BOX 1-1**
**CCAS Recommendations for Accelerator-Based Explosives-Detection Technologies**

**Recommendation 4-1. Do not consider accelerator-based technologies to have promise for deployment as a primary screening procedure for checked baggage inspection.** Any screening procedure relying on an accelerator cannot compete with available technologies on either cost or practicality bases.

**Recommendation 4-2. Do not fund any large accelerator-based hardware development projects.** Combinations of experimental work with existing laboratory equipment, mathematical modeling, and simulation can better define the potential of the nuclear technologies without the expense or time required to design and build new hardware.

Source: NRC, 1997.

sistent with the FAA EDS specification (Chmelik et al., 1997). In spite of this level of performance, the CCAS did not believe such a system should be fielded or even that optimal fielding configurations should be investigated. The committee concluded that testing should be conducted with existing laboratory systems and smaller pixel sizes to determine the detection limits of PFNTS for Class A explosives. Only if this testing demonstrated that the detection performance[6] could be substantially improved should the FAA investigate the potential application of this technology.

The CCAS observed that the principal advantage of PFNTS may be its potential for resolving alarms raised by a lower-cost, high-resolution, image-based explosives-detection device or system (NRC, 1997). In order to demonstrate this potential, more will need to be done than determining the detection performance on a cluttered bag set consistent with the bag set used in the current FAA EDS certification testing. The PFNTS false alarm performance would have to be determined for a set of bags that set off the alarms of the best conventional (non-nuclear) EDSs.

In 1994, the FAA awarded Tensor Technology[7] a two-year grant to build a multidimensional neutron radiometer (MDNR)[8] airline security system, including transporting the system to a nuclear accelerator and testing it to determine its sensitivity for detecting explosives concealed in suitcases (Tensor Technology, 1998a). In these tests, the detection performance of the MDNR showed promise for meeting the probability of detection ($P_d$) required for FAA certification for all but one category of explosives. Considering these test results and the CCAS recommendations listed in Box 1-1, the FAA awarded Tensor a six-month cooperative agreement grant to present the company's evaluation of PFNTS compared to other, currently available technologies for the primary screening of passenger baggage for explosives and for the screening of cargo in airports. Tensor was asked to include the following points in its evaluation of PFNTS:

- operational requirements, including size, weight, power requirements, cooling, and other utility requirements, as well as placement options (bag room or a separate building away from public airport buildings)
- operational aspects, including operator training requirements, alarm resolution procedures, impact on air carrier operations, access to baggage, and baggage flow
- performance levels, including expected and measured detection and false alarm rates for all relevant types of explosives
- availability of equipment, including mean time between failures, mean time to repair, required parts inventory and lead time to obtain parts, and maintenance costs
- safety concerns, including radiation safety and monitoring, U.S. Nuclear Regulatory Commission licensing, and radiation shielding
- costs, including initial unit cost and installation costs, site preparation costs, costs of modifying belt lines and baggage-handling equipment and support equipment, and operational costs for utilities and environmental control
- cost to finish development compared to technical risk

explosives in this report. Class B explosives include all other explosives in the FAA's certification test set. A detailed description of Class A and Class B explosives is not available in this report due to the sensitive nature of this information. Specific questions regarding the performance of PFNTS should be addressed to the FAA.

[6] In this context, *substantial improvement* involves achieving a high probability of detection while maintaining a low probability of false alarms.

[7] Tensor Technology, Inc., Madison, Alabama.

[8] Tensor Technology refers to their PFNTS-based explosives-detection device as a multidimensional neutron radiometer (MDNR) airline security system.

**BOX 1-2**
**Statement of Task for the Panel on Assessment of the Pulsed Fast Neutron Transmission Spectroscopy for Aviation Security**

The panel will evaluate the potential for pulsed fast neutron transmission spectroscopy for primary passenger baggage screening or cargo screening compared to currently available x-ray based computed tomography systems. To accomplish this the panel will:

- review and assess Tensor Technology's report, which is expected to address both technical and operational capabilities and projected capabilities of PFNTS for primary passenger baggage screening and cargo screening
- review the laboratory-demonstrated explosives-detection performance of PFNTS
- compare demonstrated and projected capabilities with those of currently available x-ray-based computed tomography systems
- evaluate the potential that end users will prefer a PFNTS-based system to currently available x-ray-based computed tomography systems
- outline any key assumptions that would be required to envision the use of PFNTS in airports and, if appropriate, recommend strategies to confirm these assumptions
- develop guidelines for the FAA to follow to determine the feasibility of PFNTS technology for use as explosives-detection equipment in airports

The FAA requested that the NRC review and evaluate Tensor Technology's assessment of PFNTS in light of the previous recommendations of the CCAS (see Box 1-1) and in light of technical developments since the committee's second interim report. In response to the FAA's request, the NRC convened the Panel on Assessment of the Practicality of Pulsed Fast Neutron Transmission Spectroscopy for Aviation Security under the auspices of the CCAS. The panel was charged with evaluating the practicality of PFNTS for the primary screening of passenger baggage or for cargo screening, as compared to currently available x-ray-based CT systems (see Box 1-2).

## ORGANIZATION OF THIS REPORT

This report assesses the practicality of PFNTS for aviation security. The principle of bulk explosives detection is discussed in Chapter 2. Chapter 3 contains an assessment of laboratory test results for PFNTS, and Chapter 4 discusses the laboratory and operational performance characteristics of the FAA-certified x-ray CT-based InVision CTX-5000.[9] Tensor's report is reviewed in Chapter 5, and the panel's comparison of PFNTS and the CTX-5000 is presented in Chapter 6. The panel's conclusions and recommendations are presented in Chapter 7.

[9] InVision Technologies, Newark, California.

# 2

# Principle of Bulk Explosives Detection

In this chapter, the principles behind the technologies developed for detecting explosives concealed in baggage are discussed. The two major categories are x-ray-based technologies and neutron transmission-based technologies. The discussion of x-ray-based technologies covers those currently deployed in airports, as well as some modified implementations. The discussion of neutron-based technologies is limited to PFNTS—the focus of this study.

## X-RAY-BASED TECHNOLOGIES

Many of the explosives-detection technologies that are based on x-ray techniques measure the x-ray attenuation of the materials that make up the baggage. Attenuation is a function of energy, density, and average atomic number. Because x-rays interact primarily with electrons, the attenuation coefficient is strongly correlated with the electron density of the material under investigation (NRC, 1998).

The mechanisms primarily responsible for x-ray attenuation in materials at the x-ray energy ranges typically used by explosives-detection equipment are Compton scattering and photoelectric absorption. The photoelectric effect results in x-ray absorption, whereas Compton scattering merely scatters x-rays, altering the path and energy of the scattered photons (x-rays). The significance of the photoelectric effect is greater for materials composed of elements with a high atomic number ($Z$), such as metals or other inorganic materials. However, this cross section drops off rapidly (i.e., the attenuation caused by the photoelectric effect becomes less relevant) with increasing x-ray energy. For organic materials (low $Z$), Compton scattering is the dominant x-ray attenuation process. The attenuation cross section caused by Compton scattering varies less with x-ray energy than does the attenuation cross section caused by the photoelectric effect. Materials can be distinguished from each another based on the relative importance of the photoelectric and Compton cross sections. For example, inorganic materials can be identified by rapidly changing x-ray attenuation with changing x-ray energy, whereas organic materials display a more subtle change (NRC, 1998).

Multi-energy x-ray-based detection equipment suitable for distinguishing organic from inorganic materials and for measuring densities semiquantitatively has been developed. Combining the measurement of transmission and backscatter x-rays improves the detection of light (low $Z$) elements as they are found in explosives; however, it does not specifically identify explosives. Dual-energy CT (computed tomography) is capable of providing geometrical information, as well as information pertaining to both the physical density and the effective atomic number of a material. Although the effective atomic number is not enough to completely characterize a material, it does provide discrimination capability above and beyond characterization by a physical density metric alone.

A common imaging method, x-ray radiography (or projection imaging) is a collection of x-ray attenuation line integrals over two dimensions. This method does not resolve the third dimension along the incident x-ray direction. CT adds the capability of visually displaying the physical appearance of the materials in question from all three dimensions. Reconstructing two-dimensional cross-sectional images (tomographs) and then full three-dimensional volumes can greatly improve the detection of explosive threats by identifying certain shapes or patterns, such as wires, batteries, or detonators, as well as by measuring the volume of the suspect material. The additional geometrical information supplements the material x-ray attenuation information and results in a more specific discrimination of explosive materials.

To date, four EDSs have been certified by the FAA, all of which are x-ray CT-based systems. Three are manufactured by InVision Technologies, Inc.: the CTX-5000, CTX-5000 SP, and CTX5500 DS; a fourth system, the 3DX-6000, which recently passed the FAA certification test, was developed by L3 Communications, Inc.

Other x-ray-based methods using high-energy photons (between 10 and 30 MeV) have been discussed in the literature (Hussein, 1992; Gozani, 1988). Because of low-reaction cross sections, they require high x-ray flux rates produced by powerful accelerators that may not be suitable for airport use. These high-energy techniques are based on photon interactions with the nuclear properties of nitrogen, carbon, and oxygen. They provide spatial information, but because of the small-reaction cross sections, it is difficult to distinguish between elements.

## PULSED FAST NEUTRON TRANSMISSION SPECTROSCOPY

In the PFNTS method, a collimated broad-energy (0.5–8 MeV) or "white" neutron beam is passed through the bag and the energy-dependent neutron transmission measured. By comparing the energy-dependent attenuation of the source neutron spectrum, the ratios of hydrogen, oxygen, carbon, and nitrogen can be integrated over a path (line) in the bag, and multiple lines can be used to produce a radiographic image (Overley, 1985). A fictitious element ($X$) with a smooth energy-dependent cross section is often included to normalize the transmitted number density (Overley et al., 1997). $X$ is intended to represent a smooth neutron attenuation, which can be attributed to elements not specifically represented in hydrogen, oxygen, carbon, and nitrogen decomposition. For every pixel in the target, the energy dependence in the transmitted neutron spectrum is used to determine the relative amounts of these five elements. Figures 2-1a and 2-1b (Chmelik et al., 1997) show how projections in these five dimensions can be used to distinguish the presence of an explosive and often (~ 72 percent of the time [Lefevre and Overley, 1998]) to identify the type of explosive. The points in Figure 2-1a and 2-1b represent 38,000 measurements from actual airline suitcases, with and without explosives. Because of the scatter in the plotted points for explosive and nonexplosive paths, it is difficult to apply projected-path nitrogen-only detection schemes effectively. Because of the outliers in the distribution of "nonexplosive" black points, it is difficult to eliminate false alarms without affecting the probability of detecting the explosive.

In the basic PFNTS method, a five-dimensional representation of the elemental composition and the spatial distributions of "potentially explosive" adjacent pixels are used to support the detection algorithm. Various algorithms can be used to reduce the five-dimensional elemental information to an "explosive potential" for a single pixel. Variations in the detection algorithm can also increase the base set beyond the nominal five elements. The Tensor algorithm includes another element ($Y$), which is changed from element to element within a specified set of cross sections during a regression calculation until a best fit is obtained (Tensor Technology, 1998b). Different spatial correlation algorithms can be used to reduce the map of "explosive potential" metrics to a yes or no decision on the presence of an explosive in the test article. The University of Oregon refers to its detection algorithm as a "B-matrix" and bases the "explosive potential" metric on a comparison with the explosive/nonexplosive probability observed in a simulation database. Separate B-matrices are maintained for each explosive class the algorithm is designed to detect. Tensor uses a neural network trained on a set of explosive and nonexplosive bags.

PFNTS requires the use of a tightly bunched, pulsed neutron source. Time-of-flight is used to determine the energy of the transmitted neutrons. The temporal width of the initial neutron pulse and the time resolution of the time-of-flight measurement limit the energy resolution of the transmitted spectrum. Flight paths of 4 to 10 m (13–33 ft) are commonly used. Figure 2-2 shows the total cross section for neutrons on $^{16}O$, $^{14}N$, and $^{12}C$. The narrow peaks in the interaction cross sections do not appear in a typical PFNTS measurement because they are smeared out by the energy resolution of the detectors (Miller and Makky, 1993). Thus, the element identification depends on the broad energy-dependent structures in the cross sections, rather than on the narrow resonances.

Because this method measures the energy-dependent neutron attenuation, it is critical that a broad energy neutron source be used. This rules out $^{2}H(^{2}H,n)^{3}He$ (referred to as a deuterium-deuterium or DD reaction) and $^{2}H(^{3}H,n)^{4}He$ (referred to as a deuterium-tritium or DT reaction) sources, which have a monoenergetic or restricted energy range. In order to get a reasonable neutron flux for high-energy neutrons (up to 8 MeV), accelerators are generally required. $^{9}Be(d,n)^{10}B$ or $^{9}Be(p,n)^{9}B$ reactions are candidate neutron sources (Micklich et al., 1996). Accelerators that can exploit these neutron-producing reactions need a high current (~10 mA time-averaged current, 1-ns pulse width, and 1-ms repetition frequency) and a high-energy deuteron source (> 4 MeV). Most laboratory experiments with PFNTS have used a deuteron linear accelerator and the $^{9}Be(d,n)^{10}B$ reaction.

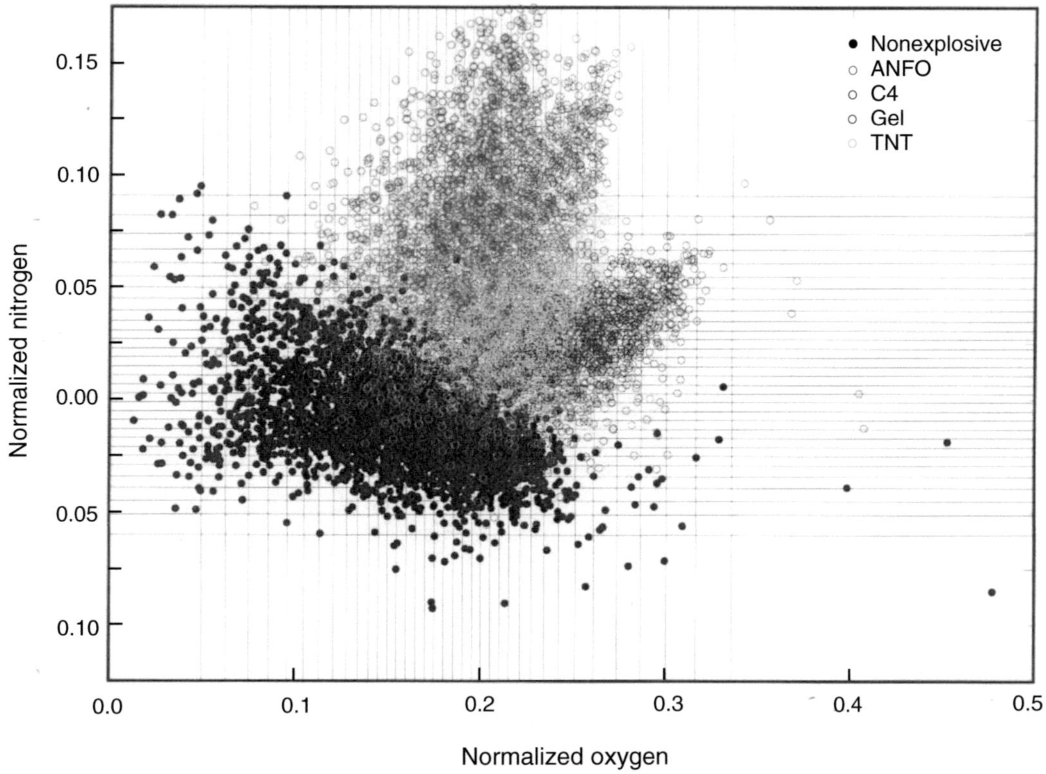


FIGURE 2-1a Normalized nitrogen and oxygen distributions determined by PFNTS from the contents of suitcases, with and without explosives. Source: University of Oregon.

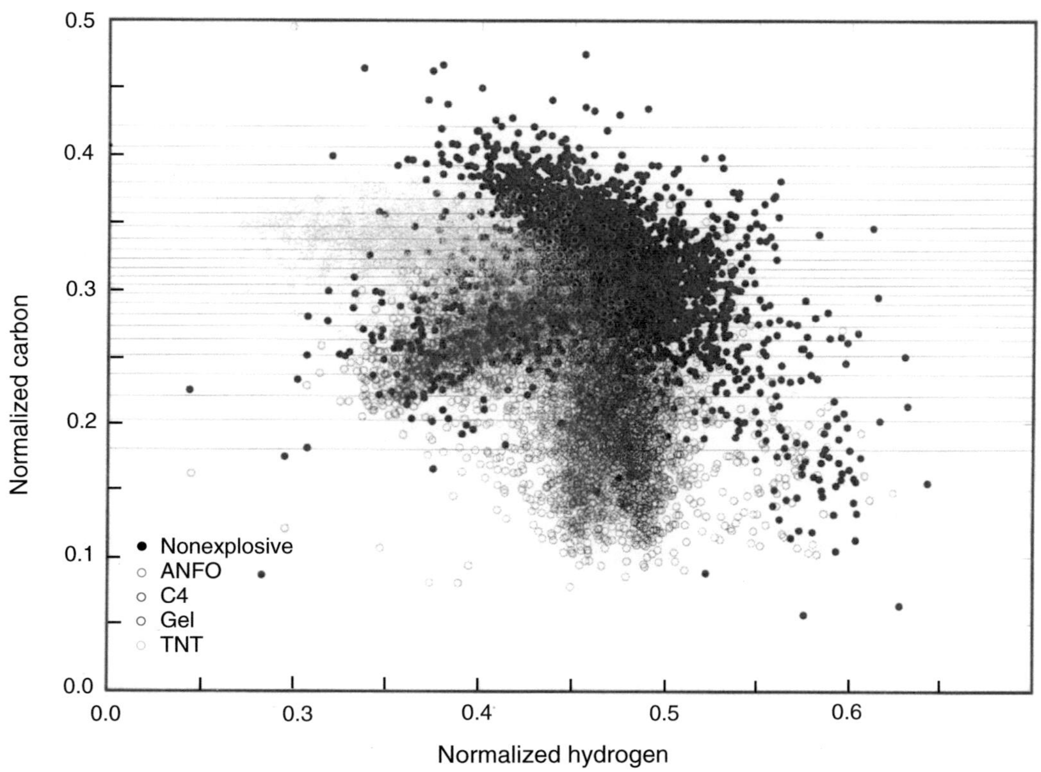


FIGURE 2-1b Normalized carbon and hydrogen distributions determined by PFNTS from the contents of suitcases, with and without explosives. Source: University of Oregon.

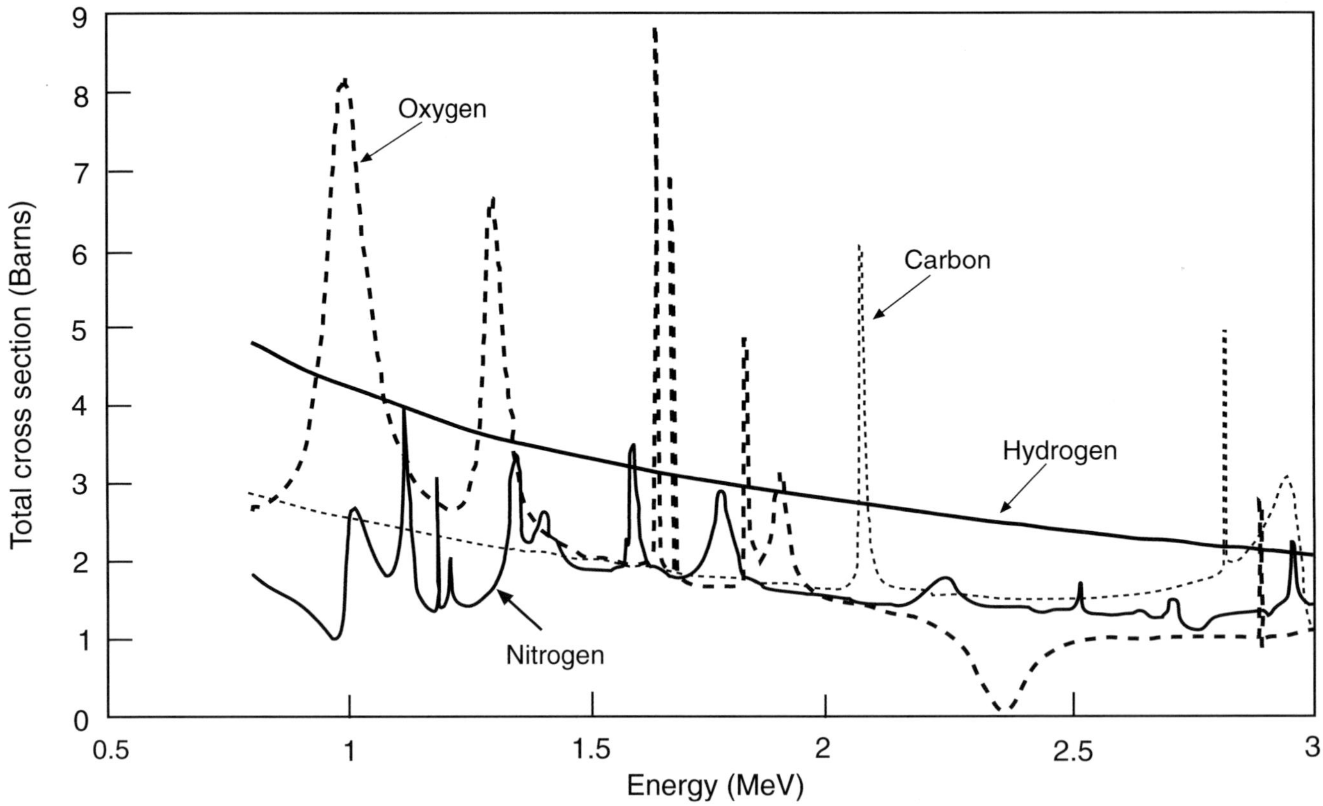


FIGURE 2-2 Total cross section of hydrogen, carbon, nitrogen, and oxygen as a function of energy. Source: Miller and Makky, 1993.

# 3

# Laboratory Tests of Pulsed Fast Neutron Transmission Spectroscopy

PFNTS technology was tested for its capability to meet both the $P_d$ (probability of detection) and the probability of a false alarm ($P_{fa}$) specified in the EDS certification standard for the range of explosive classes. The FAA conducted blind tests on two PFNTS-based explosives-detection devices, one developed by the University of Oregon and the other by Tensor Technology. The blind tests measured detection performance (coupled $P_d$ and $P_{fa}$) but not bag throughput. The results described below can be used to set the demonstrated PFNTS detection performance level.

## UNIVERSITY OF OREGON BLIND TESTS

The FAA conducted a series of blind tests (Chmelik et al., 1997) at the University of Oregon in September 1996 using the PFNTS B-matrix approach (Lefevre and Overley, 1998) and two spatial correlation algorithms, a *contiguous-pixel test* and a *shape test*. The software used to implement the B-matrix detection algorithm along with the blind test database have been archived (Lefevre, 1998). The University of Oregon detection system examined 3.2 x 3.2 $cm^2$ (1.3 x 1.3 $in.^2$) pixels of a suitcase. The test involved 134 suitcases and eight different nitrogen-based explosives. The FAA tester placed explosives in 75 of the bags. In six instances, two or more different explosives were placed in a single bag. Attempts were made to elicit false alarms by placing a variety of available materials in the bags. No attempt was made to combine materials along the neutron path to reproduce the elemental densities or ratios expected of explosives in cluttered bags. Attempts were made to conceal explosives by placing them in iron or aluminum pipes, behind books, wood, and other objects, and in radios or video cassettes.

Attempts to conceal explosives were mostly unsuccessful. In the initial blind test, using the automated *contiguous-pixel test,* 67 of the 75 explosive-containing bags were correctly identified, for a $P_d$ of 89.3 percent. Eight of the 59 bags that did not contain explosives did set off an alarm, for a $P_{fa}$ of 13.6 percent. Using the automated *shape test*, the presence of explosives in 70 of the 75 explosive-containing bags was detected, for a $P_d$ of 93 percent; seven of the 59 bags that did not contain explosives set off alarms, for a $P_{fa}$ of 11.8 percent. With operator intervention, the *shape test* operator, relying primarily on past experience, correctly identified 71 explosive-containing bags out of 75. However, in this test only 46 of the 59 bags that did not contain explosives were correctly identified. These results correspond to a $P_d$ of 94.7 percent and a $P_{fa}$ of 22 percent. Blind tests conducted at the University of Oregon, unlike EDS certification tests, distinguished between "true detections" and "false detections," that is, false alarms were registered if the explosive was not located in the portion of the bag identified by the algorithm. None of the 59 benign bags registered a false alarm. Most of the false alarms were produced by large areas of only slightly elevated explosive probabilities, which corresponded to the location of large books or blocks of wood. Table 3-1 summarizes the detection performance for the various algorithms used during the blind tests.

In four of the six cases where multiple explosives were placed in the bag, the system was able to detect the presence of each explosive. In the other two cases involving multiple explosives, the explosives were in close proximity and were identified as only one explosive. In 38 of the 75 explosives-

TABLE 3-1 Performance of the University of Oregon Explosives-Detection Algorithm in Blind Tests

| Detection Algorithm | $P_d^a$ | $P_{fa}$ |
|---|---|---|
| Contiguous pixel | 89.3% | 13.6% |
| Shape test | 93% | 11.8% |
| Operator intervention | 94.7% | 22% |
| Post-test algorithm adjustment | 93.3% | 4.5% |

[a] $P_d$ indicates a "true detection," the correct identification of the region of the bag containing the explosive.

TABLE 3-2 Performance of the Tensor Explosives-Detection Algorithm in Blind Tests

| Detection algorithm | Explosives Class[a] | $P_d$ | $P_{fa}$ |
|---|---|---|---|
| Operator assisted | B | 94.3% | 25% |
| Operator assisted | A | 51.8% | 25% |
| Operator assisted | A-1 | 83.3% | 25% |
| Operator assisted | A-2 | 43.2% | 25% |
| Automated neural net (with 10 scans eliminated because of possible interference) | all classes | 88% | 24% |
| Automated neural net detection on loose cargo when trained on bags | all classes | 70% | see note |
| Automated neural net detection on loose cargo when trained on bags | A | 40% | see note |

[a]Explosive classes A-1 and A-2 are subclasses of explosive class A.
Note: Although no false alarms were recorded, only 4 of the 60 cargo scans did not contain an explosive. Therefore, $P_{fa}$ is not statistically meaningful in this case.

containing bags, the PFNTS detection algorithm not only indicated the presence of an explosive but also successfully identified the explosives class. In another 16 cases, the actual test explosive corresponded with the second most likely explosives class predicted by the detection algorithm.

When the blind test pixel data were used to modify the B-matrix method, the error rate was significantly reduced. This post-processed blind test analysis missed five out of the 75 explosives and produced no false alarms in the 59 benign bags. If the six cases where the wrong region of the bag was identified as containing an explosive (when in fact there was an explosive in a different region of the bag) were considered to be false alarms, the $P_d$ was 93.3 percent, and the $P_{fa}$ was 4.5 percent (Algorithm 4 in Table 3-1). The performance for this algorithm must be treated as an indicator of the "potential" detection performance ($P_d$ coupled with $P_{fa}$) for the PFNTS and not as a valid blind test result because post-processing of blind test data using detection algorithms optimized to the test data can produce misleading results.

The explosives the University of Oregon *expected* would be tested, which were specified in the "FNTS Developmental Test Plan," had a defined configuration. However, different than expected explosive configuraitons were used in the actual blind tests. The test final report (Overley, 1998) indicates that seven out of the eight missed explosives involved smaller amounts of explosive than the University of Oregon researchers expected to detect.

## TENSOR TECHNOLOGY BLIND TESTS

In blind tests at Tensor Technology, Inc., in late September 1997, 150 suitcases were scanned at two angles, 0° (broadside) and 60° in azimuth (Gibson et al., 1997; Tensor Technology, 1998a). The initial determination was performed by an operator and was "somewhat subjective" but was "based on a combination of explosive size, general level of attenuation, atomic number density analysis, and the neural net analysis." Table 3-2 shows the Tensor detection performance. The $P_d$ was 94.3 percent for Class B explosives and 51.8 percent for Class A explosives (83.33 percent for Class A-1 and 43.18 percent for Class A-2). The $P_{fa}$ was 25 percent. There was no significant difference in detection performance between the two scan angles.

An automated explosive detection algorithm is under development using the regression neural net analysis. A preliminary version incorporating the neutron attenuation magnitude and the nitrogen content was used to analyze the blind test scans. The details of the blind test results from this automated algorithm are not clear from Tensor's final report to the FAA (Tensor Technology, 1998b). However, the report suggests that the operator response was somewhat better than the initial automated response. Some questions have been raised about 10 test scans in which, because of attempts to limit interference from a table used to hold the test articles, the explosives may not have been in the detector array's field of view. If these 10 scans are omitted, the automated response algorithm had an overall $P_d$ (for Class A and Class B) of 88 percent and a $P_{fa}$ of 24 percent.

The Tensor blind tests included tests of the system's potential for screening loose cargo consisting of boxes with varied contents. Thirty loose cargo items were scanned twice each for a total of 60 scans. Explosives appeared 56 times during the testing, including four cases in which two explosives were detected in a single cargo item. A neural network detection algorithm trained on data acquired from luggage (not loose cargo) was used. The overall $P_d$ was 70 percent. The $P_d$ for Class A explosives was only 40 percent. No false alarms were recorded during the cargo scanning.

Some of the loose cargo containers were fairly dense. Six of the 17 missed explosives were in containers with a neutron transmission fraction of less than 0.003, which is comparable to noise levels of background scattered neutrons during screening. In other words, transmitted neutrons could not be differentiated from background noise caused by scattered neutrons at this low transmission level.

## DETECTION OF CLASS A EXPLOSIVES

The PFNTS method has problems detecting Class A explosives (required by the FAA for certification). This may be an intrinsic limitation for explosives-detection approaches that use a large pixel size (> 0.2 cm [> 0.08 in.]). Figures 3-1 and 3-2 show the resolution available to the detection algorithms used in the blind tests. Except for Class A explosives, the PFNTS detection performance ($P_{fa}$ and $P_d$) has the potential to meet the FAA's EDS certification requirements. Because the Tensor and University of Oregon tests used very different detection algorithms and because both provided good detection levels for Class B explosives, the PFNTS elemental densities seem to provide a very robust set of measurements for explosives detection. Refinements in the detection algorithm or coupling with other technologies may be necessary, however, to overcome the limitations of PFNTS for the detection of Class A explosives. Once this limitation has been better quantified, it should be possible to determine the potential role for PFNTS in commercial aviation security.

The unreliable detection of Class A explosives is a serious deficiency of PFNTS, particularly for Class A-1 explosives. However, according to reports by the University of Oregon (Overley, 1998), some subclasses of Class A explosives can be reliably detected. Approaches to improving the capability of PFNTS to detect Class A explosives are listed below:

- reducing the pixel dimensions at the bag location
- using multiple scans at different angles and a tomographic analysis of the data
- improving spatial correlation algorithms so that large areas with a low probability of being an explosive are identified as an explosive

Determining the efficacy of these approaches will require further laboratory testing. Some of these approaches (e.g., tomographic analysis) will reduce the bag throughput rate compared to a single-view radiographic method.

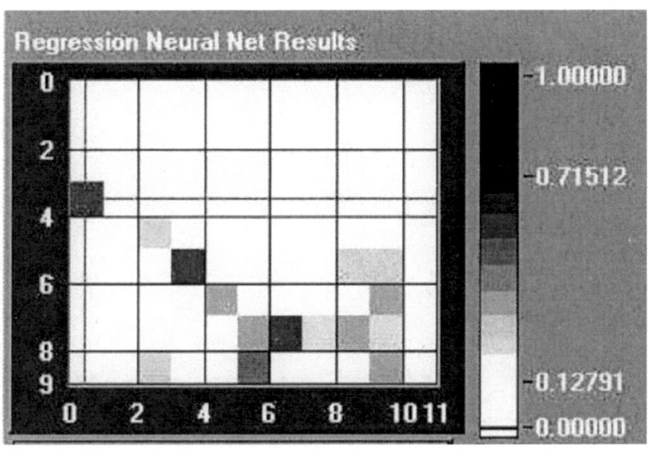


FIGURE 3-1 Neural net values during Tensor blind testing for a slurry sample at an angle in a suitcase. Source: Tensor Technology, 1998a.

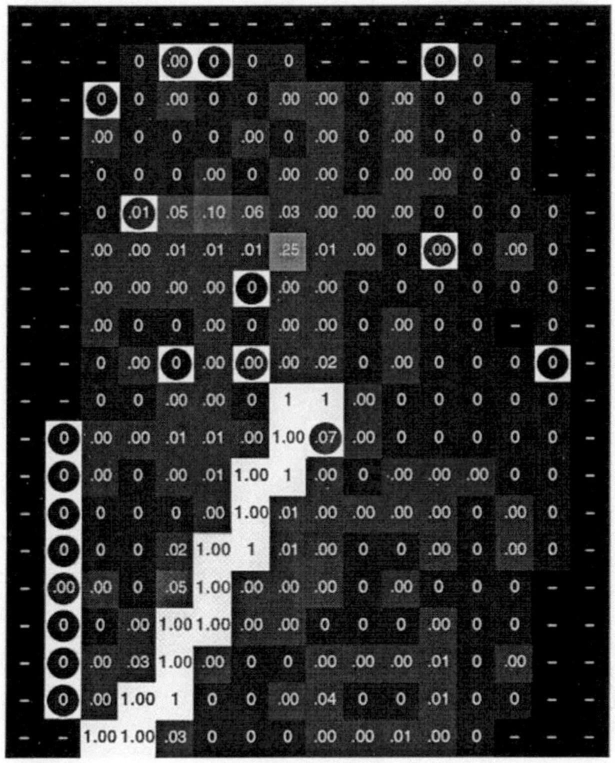


FIGURE 3-2 Gray-scale maps from B-matrix during University of Oregon blind tests of a bag containing an explosive in an iron pipe sloping up to the right. Source: University of Oregon.

## ASSESSMENT OF DETECTION PERFORMANCE

The post-processing of blind test data using detection algorithms optimized to the test data can produce misleading results. For example, if the detection algorithm is optimized to specific test data, it is possible that when new data are analyzed (i.e., from a new set of baggage), the algorithm will not perform as well. Although the 4.5 percent $P_{fa}$ and 93.3 percent $P_d$ described above for post-processed data from the University of Oregon PFNTS tests indicate a detection performance level that *may* exceed the explosives-detection potential of x-ray-based CT approaches, these data must be treated with a healthy skepticism. Furthermore, the low $P_{fa}$ attained during the University of Oregon tests—with a 16-element linear array detector—may not hold for a full two-dimensional array. The University of Oregon acknowledges that changing to a two-dimensional array detector could degrade the performance of PFNTS for the following reasons (Lefevre, 1998):

- As the detector geometry is expanded, the background noise level becomes much more severe (i.e, more scattered neutrons enter the scanning area).

- More collimator area is visible to each detector, which affects the line shapes for monoenergetic neutrons by increasing long flight-time tails.
- Neutron in-scattering from luggage items increases.
- Detector *cross talk* increases.

For all of these reasons, the panel cannot confidently state that the PFNTS system has the *potential* for a very low false alarm rate under conditions that meet the FAA's required throughput rate. The problems listed above will have to be addressed through more laboratory experimentation or detailed radiation transport modeling. The Tensor two-dimensional 99-element array detector exhibited a much higher false alarm rate during blind testing (compared to the University of Oregon tests), which may be associated with the background from a different neutron source configuration or the use of a different explosives-detection algorithm rather than intrinsic characteristics of the two-dimensional detector array.

# 4

# Baseline Characteristics of Explosives-Detection Systems Based on X-ray-Computed Tomography

A legitimate comparison of x-ray CT-based EDSs and PFNTS-based explosives-detection techniques must be based on their baseline performance. Laboratory test results for PFNTS were presented in Chapter 3. In this chapter, both the laboratory performance and the operational performance of the FAA-certified CTX-5000 SP are reviewed. To date, most of the performance data on deployed explosives-detection equipment has been generated from tests conducted at the FAA Technical Center. However, operational (field) test data on false alarm rates are also reviewed in this chapter.

## TEST DATA FROM THE FAA TECHNICAL CENTER

Most of the performance data available for x-ray CT-based EDSs are from certification tests of the InVision CTX-5000 SP in 1996. Data from the certification tests of the recently certified InVision CTX-5500 DS and L3 Communications 3DX-6000 are not available yet but should be included in future analyses. The InVision CTX-5500 DS has been certified for two different inspection modes: SURE98 Mode, which has a lower false alarm rate than the CTX-5000 SP but a lower throughput rate, and CERT98 Mode, which has a similar false alarm rate to the CTX-5000 SP but a much higher throughput rate. The panel's analysis of performance data focuses on the CTX-5000 SP, although performance data for the CTX-5500 DS are also presented for reference. Table 4-1 shows the performance factors for the CTX-5000 SP and CTX-5500 DS, including the $P_d$, $P_{fa}$, and the bag throughput rate. Because the actual $P_d$ and $P_{fa}$ numbers are classified,[1] $P_d$ is given as a percentage of the overall $P_d$ required for certification ($X$), and $P_{fa}$ is given as a percentage of the $P_{fa}$ required for certification ($Y$).

## OPERATIONAL DEMONSTRATION

In 1995, the FAA initiated the Airport Operational Demonstration Project (FAA, 1995) to determine the operational performance of the InVision CTX-5000 SP in the field as compared to its performance in certification tests. Three test sites were selected for the project: San Francisco International Airport (United Airlines); Atlanta Hartsfield International Airport (Delta Airlines); and Manila International Airport (Northwest Airlines).

[1] The actual values required for certification are recorded in classified FAA documents (FAA, 1992).

TABLE 4-1 Performance Test Results for the InVision CTX-5000 SP and CTX-5500 DS

| Equipment | Detection Rate | | False Alarm Rate | | Throughput |
|---|---|---|---|---|---|
| | Sample Size | $P_d$ (% $X$)[a] | Sample Size | $P_{fa}$ (% $Y$)[a] | (bags/hour) |
| CTX-5000 SP | 600[b] | 106 | 1,000[c] | 90 | 245 |
| CTX-5500 DS/SURE98 mode | 600[b] | 106 | 1,000[c] | 60 | 264 |
| CTX-5500 DS/CERT98 mode | 600[b] | 100 | 1,000[c] | 95 | 362 |

[a]Percentage of the classified value required for certification.
[b]A total of 150 bags were used containing improvised explosive devices (25 samples of six explosives, detonators, timers, and wires) and four different orientations to get a sample size of 600.
[c]The 1,000 bags used to acquire the $P_{fa}$ differed from the 600 bags used to acquire the $P_d$. The 1,000-bag set was also used to obtain throughput rates.

Two InVision CTX-5000s were installed in Atlanta and one each in San Francisco and Manila. The demonstration project included four open tests and one blind (red team) test using improvised explosive devices to determine $P_d$. The $P_{fa}$ was acquired routinely throughout the project on real passenger bags. Only the data from the San Francisco and Atlanta operational deployments have been documented in final reports (FAA, 1997a, 1997b, 1997c).

Some of the performance data for the CTX-5000 SP installed at San Francisco International Airport are given in Table 4-2. The automated explosives-detection capability of the CTX-5000 SP during open testing was about the same as the capability measured at the FAA Technical Center. However, operator intervention to resolve alarms measurably reduced $P_d$. This tendency was also observed during blind testing at San Francisco and Atlanta (FAA, 1997a, 1997b). During the operational demonstration project at San Francisco, the automated $P_{fa}$ was 113 percent to 150 percent higher than the certification standard. Operational data reviewed by the inspector general of the U.S. Department of Transportation suggested that false alarm rates were as much as 169 percent higher than the certification standard. Data from the first of the four open tests show an average of 50 seconds for alarm resolution using the CTX-5000 SP. Although the resolution time was lower during subsequent tests, the combination of a high $P_{fa}$ and a long alarm resolution time can have a significant impact on throughput rates. In fact, it was determined to be the limiting factor for throughput rate.

Although the CTX-5000 SP exhibited a significantly higher automated (i.e., without operator intervention) $P_{fa}$ in operation at airports than during certification testing, the image produced by the system is used by trained operators to assist in resolving the alarms. Thus, the actual $P_{fa}$ (with operator intervention) is much lower than the automated $P_{fa}$ measured during certification testing. Nevertheless, the time required for alarm resolution is a factor that must be taken into consideration when evaluating the performance of this EDS.

TABLE 4-2 Summary of Open Testing of CTX-5000 SP at San Francisco International Airport

| Tests | Sample Size | $P_d$ (% $X$) | $P_{fa}$ (% $Y$) [a] |
|---|---|---|---|
| Machine (automated) | 131 | 102 | 150 |
| Machine + operator | 131 | 89 | 5 |

[a] $P_{fa}$ was obtained by measuring the $P_{fa}$ for regular passenger baggage.

The FAA's Security Equipment Integrated Product Team (SEIPT) has deployed more than 40 InVision CTX-5000 SP EDSs since 1997. Thus, the FAA now has an outstanding opportunity to collect performance data (e.g., $P_d$ and $P_{fa}$) as well as operational performance data (e.g., down time, mean time between failure, and mean time to repair). Unfortunately, this information was not available to the panel, which substantially compromised the panel's ability to assess the performance of deployed equipment and to make comparisons between deployed equipment and alternative technologies, such as PFNTS. Credible future assessments of the performance of deployed explosives-detection equipment will require both performance and operational data.

# 5

# Tensor Technology Report on the Multidimensional Neutron Radiometer Airline Security System

A specific item in the Statement of Task (Box 1-2) for this panel was to review the Tensor Technology report, *Advanced Studies on the Multi-Dimensional Neutron Radiometer (MDNR) Airline Security System,* (Tensor Technology, 1998a). This chapter reviews the details of the proposed MDNR. The basic MDNR design is a reasonable baseline conceptual design for assessing the airport implementation and acceptability of the PFNTS system.

## TECHNICAL CAPABILITIES AND PHYSICAL ATTRIBUTES

The panel's critique is based on the MDNR design presented in the Tensor report supplemented by discussions with representatives of Tensor. The MDNR conceptual design is shown in Figure 5-1. Tensor has proposed a very innovative approach to the PFNTS neutron source, using a cyclotron rather than a linear accelerator to reduce the footprint of the explosives-detection device. Although this approach has some advantages, it also raises some questions because all of the previous tests on the detection capabilities of PFNTS were conducted on equipment with a linear accelerator as the neutron source. Furthermore, the cyclotron-based design still has weight and space requirements that may preclude integration into the passenger baggage line at existing airports. Although cyclotrons are used in many hospitals for the production of short-lived radioisotopes, the environmental controls and support utilities available at a hospital are different from those of a typical airport baggage makeup room.[1]

The following discussion covers important aspects of the MDNR technical capabilities. Many of the performance characteristics are summarized in Table 6-1, which compares the performance and operational attributes of MDNR and the InVision CTX-5000 SP.

### Performance Levels ($P_d$ and $P_{fa}$)

The $P_d$ and $P_{fa}$ for the cyclotron-based MDNR are not well established. Because of the lack of testing at cyclotron-based neutron sources, this review is based on performance estimates for linear accelerator-based tests performed at the University of Kentucky and the University of Oregon. In the university tests and in the proposed MDNR design, the $^{9}Be(d,n)^{10}B$ reaction was used for neutron production. One important potential difference between the cyclotron and the linear accelerator involves the energy and spatial stability of the deuteron beam. The cyclotron deuteron beam is stable to within about ± 200 keV, or about one revolution in the cyclotron. The linear accelerators have a beam homogeneity of about < 5 keV while operating at a nominal deuteron energy of 4.2 MeV. The larger spread in the deuteron energies is not expected to be a problem, but temporal variations in the deuteron energy would probably degrade performance. Any variation in the flux of deuterons delivered to the target (the beam current) must be compensated for by other beam diagnostics. Because beam current can usually be easily measured when the beam is not impinging on the beryllium target, the long-term current drift can be compensated for; nevertheless, short-term variations are a matter for concern.

The Ebco Technologies[2] cyclotron (proposed for use by Tensor in the MDNR) has a reported current stability of < 1 percent; the University of Oregon linear accelerator has a reported voltage stability of < 0.1 percent. These data, which are included in the facility documentation, are not for equivalent parameters and do not address the temporal dependence of the parameter variations. The voltage is related to the energy of the deuteron and hence affects the neutron energy

[1] The airport baggage make-up room, where baggage is prepared to be loaded onto airplanes, is one location baggage could be screened for explosives.

[2] Ebco Technologies, Inc., Richmond, British Columbia, Canada.

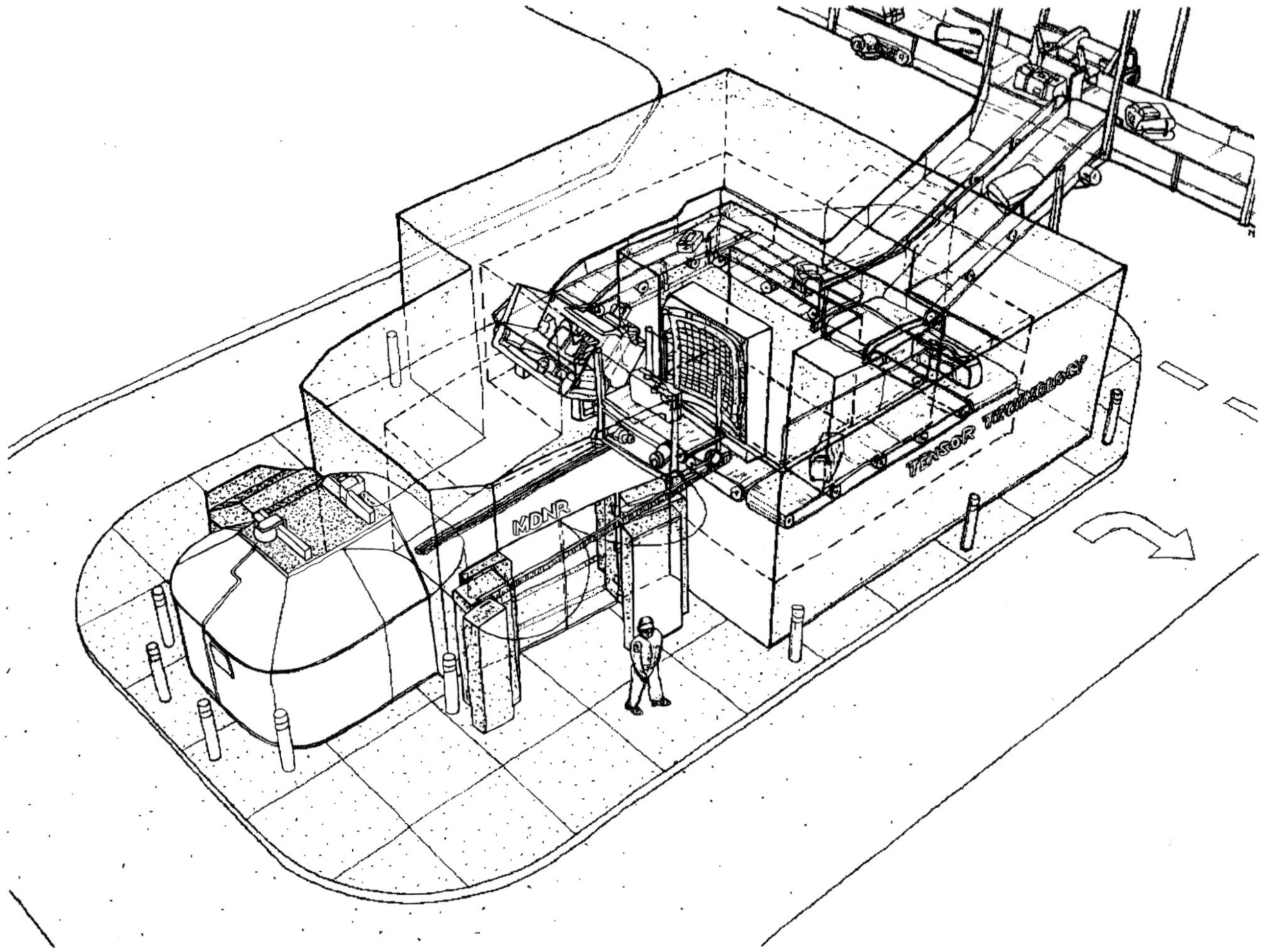


FIGURE 5-1 Artist's conception of the layout of the MDNR. Source: Tensor Technology, 1998a.

spectrum. The current is related to the number of deuterons per second impinging on the target and hence the number of neutrons impinging on the scanned bag. The panel believes the MDNR detection capability should be based on tests conducted with cyclotron neutron sources or on assurances that the short-term (less than a bag scan time, nominally ~7 seconds to support an EDS throughput of 500 bags per hour) cyclotron voltage and current variations are less than those demonstrated in laboratory tests with linear accelerators.

Based on the linear-accelerator blind tests by Tensor at the University of Kentucky (see Chapter 3) to determine the MDNR detection performance, the $P_d$ is ~95 percent for Class B explosives, ~50 percent for Class A explosives, and 0 percent for some configurations of Class A explosives. Tests at the University of Oregon yielded similar results. However, to be valid the $P_d$ of a system must be accompanied by the $P_{fa}$. The University of Kentucky tests had a $P_{fa}$ of ~25 percent. The University of Oregon tests demonstrated a $P_{fa}$ of ~12 percent using the original detection algorithm and a $P_{fa}$ of ~5 percent using post-processing analysis. The differences in the $P_{fa}$ at the two laboratories raise some concerns. Differences in $P_{fa}$ may be related to the use of an open two-dimensional detector at the University of Kentucky rather than the linear array used at the University of Oregon, to differences in the explosives-detection algorithms, or to differences in the test protocols.

In any event, these laboratory test results indicate that the MDNR would probably *not* meet the EDS detection performance requirements for Class A explosives.[3] Results indicate that the MDNR *would* meet the EDS detection requirements for Class B explosives and, indeed, for all classes of explosives in the FAA EDS specifications except Class A explosives. Based only on the current *demonstrated* performance of PFNTS for all explosive classes PFNTS is not very

[3] The Class A explosives-detection performance during certification testing depends on the distribution of configurations.

impressive and would not merit further development. Many explosives-detection systems have come close to the certification-level of detection for Class B explosives but have failed for Class A explosives and often one or two subclasses of Class B explosives. Indeed, most other explosives-detection equipment did not meet the certification-level detection requirements for Class A explosives. PFNTS has little potential for being combined with another detection technology to fill this performance gap. The only possibility for a complementary technology may be nuclear quadropole resonance, which has demonstrated very good performance for Class A-1 explosives but has not demonstrated adequate detection of Class A-2 explosives in large bag configurations.

The panel believes that the post-processed analysis from the University of Oregon tests should be validated better. If the post-processing analysis is representative of future test results (and not unduly tuned to the specific bag set) and if the $P_d$ and $P_{fa}$ results can be shown to apply to area detectors as well as linear detector arrays, then the PFNTS system performance ($P_d$ > 95 percent and $P_{fa}$ < 5 percent) may exceed the present and expected detection performance of other EDSs for Class B explosives.

## Cost

Tensor has provided details to support a cost of $2.4 million for the initial MDNR unit after the expenditure of an additional $1.8 million for engineering, assembly, and testing of the initial unit. The final units are expected to cost ~ $1.5 million if purchased in quantities of 11 or more. The installation cost is estimated to be an additional $323,000 for the vault enclosure and utility modifications and $175,000 for modifications to the baggage lines. These cost estimates were reasonably substantiated by Tensor for a conceptual design. However, the panel believes that the estimated installation cost is too low, particularly the construction costs for a shielded enclosure and modifications to the baggage line.

The panel's conclusion that the estimated installation cost is unreasonably low is based on the $1 million to $3 million cost for installation alone (i.e., without counting the purchase price) of a CTX 5000 SP fully integrated into the baggage line at some airports (the fully integrated CTX 5000 SP also requires extensive modifications to the baggage line to meet operational requirements). The average installation cost to place approximately one-third of the planned 54 CTX-5000 SP systems in airports as stand-alone units was $20,000 to $150,000 per system. These figures reflect the variability of system integration costs depending on airport design and operational constraints. The MDNR conceptual design and cost analysis are for the simplest airport baggage-line configuration. Of the four airports considered in the MDNR conceptual design study, a baggage-line integration of the MDNR system was feasible in only two. The other two would require that separate facilities be constructed to house the MDNR. Tensor's estimate of installation cost, therefore, should be considered a minimum.

When the CTX-5000 SP systems were installed, the airlines frequently (84 percent of the time) had them installed at the check-in point rather than in the baggage makeup room because it is much easier to resolve alarms when passengers are nearby. Check-in point integration would not be possible for the MDNR system. In fact, MDNR systems would have to be installed on the ground level because of their weight.

## Accelerator

One of the notable characteristics of the MDNR conceptual design is the use of a cyclotron, an accelerator with a circular ion path, rather than a linear accelerator, which was used in all previous PFNTS testing and conceptual designs. From the standpoint of integration into an airport facility baggage line, using a cyclotron is a useful selection. However, reduction in the size of the spatial footprint for the system also increases the weight of the MDNR because of the cyclotron's heavy magnets. Even though Ebco Technology's TR9D cyclotron (selected for use in the MDNR) weighs 20 tonnes (22 tons), one could argue that this is not a severe penalty compared to the 109-tonne (120-ton) shielding enclosure or the 528-tonne (581-ton) vault enclosure required for accelerator-based explosives-detection techniques.

A significant advantage of the MDNR accelerator is that it uses commercial off-the-shelf technology. Rather than a unique design, the TR9D cyclotron is basically the same as the commercial cyclotron[4] used for the production of radioisotopes. Thus, the use of a commercial off-the-shelf cyclotron would add to the manufacturability of the MDNR, as well as quality control to ensure consistent performance. A potential disadvantage of the cyclotron is that the rotating, accelerating charged particles produce much more gamma radiation than linearly accelerated particles with the same energy and charge. The MDNR overcomes this disadvantage by placing an 80-cm (32-in.)-thick borated concrete shield around the compact cyclotron, in addition to the concrete vault enclosure used to shield both linear and cyclotron accelerators.

Comparing the radiation from different types of accelerators is very complicated. With a linear accelerator, proposals for minimizing the length of the system often involve a two-story configuration, with the accelerator on one floor and the neutron flight path on another. The flooring material between the levels can be used as shielding to protect the detector from radiation. However, bending magnets are

[4]The proton version of the cyclotron, the TR19, is used for positron emission tomography radioisotope production of $^{18}F$, $^{15}O$, $^{13}N$, and $^{11}C$ at hospitals, clinics, laboratories, and radioisotope distribution centers.

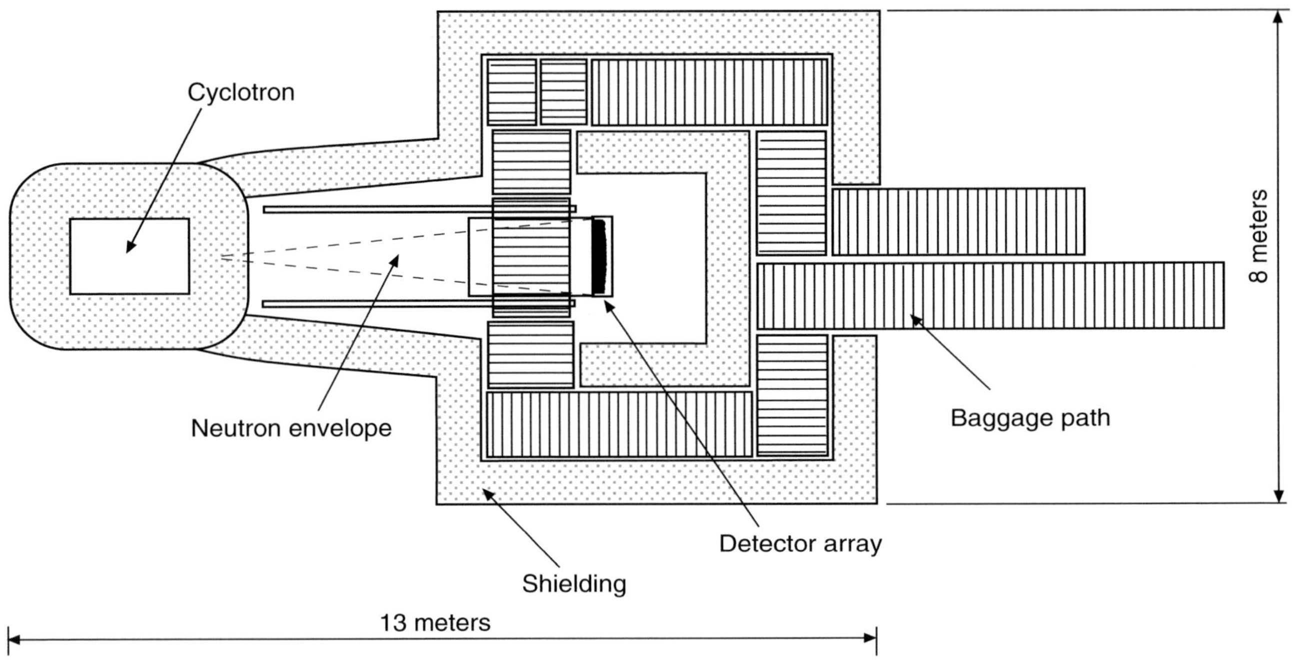


FIGURE 5-2 Possible baggage flow path for the MDNR. Source: Tensor Technology, 1998a.

often used (with linear accelerators in this configuration) on the detector floor to focus the deuteron beam on the target. These magnets are the source of intense radiation (as charged particles accelerate around the curved path), which could interfere with the system detection unless shielding is used. But, like the cyclotron magnets, the bending magnets used with a linear accelerator are a localized radiation source that can be shielded efficiently with a small volume of materials.

The potential advantages of a cyclotron over a linear accelerator—size and compact neutron source for easier shielding—may not carry over into an MDNR if it is housed in a separate facility rather than integrated into the baggage line. If a separate facility is required, size may not be a crucial factor, and the design trade-offs with linear accelerator-based systems would have to be reconsidered. The temporal variation of the beam voltage and current of a cyclotron will require more analysis and perhaps validation testing.

## Radiation Shielding

The neutron shielding on the outside of the vault enclosure in the MDNR design is intended to provide a work environment for a "nonradiation worker," that is, a radiation level of less than 2 mrem/hr with a total allowable yearly dose of less than 100 mrem. This level of shielding would simplify and reduce the cost of the radiation safety for general workers in the bag make-up area (outside of the vault enclosure) by eliminating the need for a large thermoluminescent dosimeter (TLD) program. The baseline MDNR radiation shielding consists of a 109-tonne (120-ton) shield around the cyclotron itself and a 528-tonne (581-ton) vault enclosure.

The panel was concerned about the possibility of neutron radiation streaming near the entrance to the baggage line maze in the proposed MDNR design (see Figure 5-2). According to Tensor's radiation transport calculations, the neutron radiation levels would be acceptable. However, if an 8,760-hour year is assumed, the secondary gamma radiation levels in the initial design would exceed 100 mrem per year.[5] Assuming that workers would not be in the area for more than 2,080 hours per year, Tensor concluded that the shielding was adequate. Tensor also identified some nominal design changes (e.g., a 50-cm [20-in.] borated polyethylene beam catcher) that could reduce exposure by another factor of four. The fidelity of these calculations was consistent with a conceptual design, but tensor correctly notes that a higher fidelity "engineering design study" should be done in the next phase of development to refine the estimate and uncertainty analysis of the radiation environment along the periphery of the vault enclosure.

In its calculations of radiation shielding, Tensor assumed a neutron source of $7.11 \times 10^9$ neutrons per second at the target location. Using the MDNR baseline design with a 50 mamp current, a 2 ns pulse width, and a 1.2-MHz pulse repetition rate, the total target neutron $4\pi$ production would be

[5] Based on this assumption, a worker would have to work in the proximity of the MDNR 24 hours a day 365 days a year. A more realistic assumption would be 40 hours a week, 52 weeks a year, or 2,080 hours a year.

$5.65 \times 10^{12}$ neutrons per second. This neutron level is higher by a factor of 795 than the one used in the Tensor analysis, which raises concerns about the applicability of Tensor's shielding calculations. The difference in neutron source strength reflects that Tensor's calculations were based only on the neutrons transmitted along the collimated beam and subtended by the bag. Although not obvious in the initial proposal, Tensor used a 93-cm (37-in.)-long lithium-loaded polyethylene collimator between the beryllium target (the neutron source) and the bag. Furthermore, the baseline MDNR design avoids a beam transport line and located the beryllium target at an extracted beam focal point one foot outside the cyclotron itself but within a beam portal in the 80-cm (32-in.) borated concrete shroud. Finally, the outer portion of the cyclotron is filled with borated material, so it too can function as neutron shielding. The presence of all of this neutron shielding around the target lends credence to Tensor's estimate that only the neutrons along the collimated angle subtended by the bag would scatter into the room. However, refined radiation transport calculations will be necessary to validate this estimate.

The panel concluded that the Tensor shielding is a reasonable baseline configuration for a conceptual design. Although much more detailed radiation transport calculations will be required to support an engineering-level design, the MDNR shielding mass of 637 tonnes (109 + 528) (701 tons [120 + 581]) is expected to be reasonably close to what would be determined for an engineering-level design.

## Size

The MDNR system will require an 8 x 13 m (26 x 42 ft) footprint, or an area of 104 $m^2$ (1,092 $ft^2$). This footprint includes the vault area but does not include additional support equipment or the cyclotron operations area.

Controlling the temperature and humidity of the cyclotron would require power and water for cooling. In response to questions from the panel, Tensor described five 48-cm (19-in.)-wide rack chassis cabinets that would provide this support. These racks are similar to the support equipment shown in Figure 5-3, a photograph of an Ebco TR19 cyclotron currently installed at a commercial site in Seoul, Korea.

FIGURE 5-3 Photograph of the Ebco TR19 cyclotron accelerator. This machine is the same as the TR9D cyclotron accelerator proposed for the MDNR except that it generates 19 MeV protons rather than neutrons. Source: Tensor Technology, 1998a.

The support equipment adds only about 1.2 $m^2$ (13 $ft^2$) of floor area, much of which could be inside the vault enclosure.

The cyclotron also requires some computer equipment and a cyclotron operations console, which must be located outside the neutron radiation environment of the vault enclosure. The programmable logic controllers should be located no more than 20 m (66 ft) from the other equipment.[6] According to the Ebco literature, the recommended equipment area (support equipment and operations console) requires 9.7 $m^2$ (108 $ft^2$), a slight increase over the planned MDNR footprint (Ebco Technologies, 1998). The computer equipment may, however, require some temperature and humidity controls beyond the ones normally provided in the baggage makeup room at some airports. This requirement along with the need to protect the equipment from physical damage would increase the required area.

The detector array and most of the support equipment for explosives detection could be located in the vault enclosure. However, this assumes that the explosives-detection algorithm is fully automated and precludes any operator intervention or operator-assisted alarm resolution procedures. Although the blind test results did not show significant added value from operator-assisted explosives detection, an MDNR system in an airport should have the flexibility to include operator-assisted alarm resolution. The area to support operator alarm resolution could probably overlap with the area for the cyclotron operations control.

### Weight

The panel concluded that the MDNR design provides a credible estimate of the system mass. The magnets for the cyclotron account for most of the weight of the 20-tonne (22-ton) accelerator. The radiation shielding from the cyclotron enclosure and the exposure vault area contribute another 637 tonnes (701 tons). The weight of the other system support equipment is negligible by comparison.

## OPERATIONAL CAPABILITIES

In addition to the technical capabilities and physical attributes of the MDNR, some operational capabilities will affect the practicality of its implementation into an airport environment.

### Commercial Off-the-Shelf Equipment

The use of commercial off-the-shelf equipment in the MDNR should ensure quality control, product reliability, and maintenance, as well as minimize costs. The cyclotron and associated control systems are based on commercially available systems that should continue to be available and maintainable. The MDNR neutron detector array, a customized item, is also based on readily available commercial parts.

### Radiation Safety

Radiation safety is a critical element of a PFNTS system. For safety reasons, personnel who use the system, and possibly everyone with access to the system, would require radiation safety training. Safety interlocks for access to the vault enclosure area are a part of the MDNR design. TLDs, which would be required for anyone entering the vault area, would probably not be expensive[7] but would require administrative support. Procedures for access to the radiologically controlled areas would have to be developed and posted, and a radiation safety officer would have to sweep the vault area every time it was secured to ensure that no personnel were in the area when the cyclotron was turned on. Gaseous effluent from the vault area would probably have to be sampled and monitored for radiation.

The activation of materials in passenger bags is not expected to be an issue under routine operating conditions. However, extrapolating from radiation controls used in thermal neutron analysis systems, a radiation check of each passenger bag as it exits the vault area might be required as a feature of early systems (Jones, 1990).[8]

The baggage entrance area to the vault enclosure poses special problems from a safety standpoint because an individual entering the vault area could be exposed to unacceptable levels of radiation. Passive marking of radiologically controlled areas is permitted only if all personnel have some level of radiation safety training. More stringent access controls, such as locks or monitored access, are typical of restricted areas adjacent to public areas. The bag access area for the MDNR could be controlled in several ways, but this issue would have to be addressed in an engineering-level design.

### Deactivation Procedures

The disposal of activated materials from the cyclotron shield and from the vault shielding is another issue that will have to be addressed. Even if the activation levels are low enough that the shielding materials do not require disposal as activated material, material surveys or clear bounding radiation transport calculations will be necessary to deter-

[6] According to personal communications from Ebco, 20 m (66 ft) would be desireable but is not a firm requirement.

[7] As part of a large-scale personnel monitoring program, the cost of thermoluminescent dosimeters is between \$5 and \$50 per unit with new units provided on a quarterly basis.

[8] A thermal neutron analysis system uses thermal neutrons, which have a much larger typical activation cross section than fast neutrons (> 1 MeV). Despite this difference, monitoring of passenger bags might be necessary. Scans of material exiting radiologically controlled areas are a standard feature of existing radiological facilities.

mine the activation level. An audit trail will have to be maintained for building materials to ensure that no materials[9] regulated by the Resource Conservation and Recovery Act (RCRA) are used. The cost of disposing of radioactive material, however, would not be large compared to the purchase price of the equipment. Furthermore, outside companies could be contracted to handle the disposal of radioactive material, which would relieve airport personnel of some of the administrative burden of complying with disposal regulations.

### Baggage Flow

The baggage flow through the MDNR is not given sufficient attention in the MDNR conceptual design. This issue should be addressed in much greater detail in an engineering-level design. The sharp turns in the MDNR bag flow were introduced to reduce the neutron radiation levels at the baggage entrance to the vaulted area. However, these sharp turns also increase the potential for baggage jams.

Baggage jams in the MDNR would require that the cyclotron be turned off, a qualified operator or safety officer conduct a radiation survey of the vault area, an operator enter the vault area and clear the jam, the vault area be checked for the presence of other individuals, and the safety interlocks be reengaged. This baggage-clearing procedure would take much more time than clearing a normal jam in another part of the baggage-handling system. Frequent baggage jams in the MDNR could severely compromise the baggage flow.

## CARGO INSPECTION

The Tensor report discusses the potential of PFNTS for use in scanning cargo but does not provide a conceptual design consistent with cargo-scanning requirements. The panel concluded that PFNTS technology does not have significant potential for the inspection of thick containers of hydrogenous[10] materials for explosive amounts consistent with the FAA's EDS detection requirements because neutron attenuation is too great. For thick neutron-attenuating containers, the detection capability of PFNTS would be significantly impaired. The capability of PFNTS to detect explosives concealed in thick containers has not been experimentally verified, so Tensor's predictions are based on theoretical analyses. A cost-effective experimental verification on thick containers could lead to new detection algorithms for use in highly attenuating scenarios.

The $P_d$ for explosives concealed in containers with dimensions not much larger than passenger bags would probably be similar to the $P_d$ for explosives concealed in passenger bags. Once the potential of MDNR for scanning passenger bags has been refined to encompass Class A explosives and a low $P_{fa}$ has been validated, data should be collected on scanning containerized cargo.

---

[9]Materials, such as cadmium plating on screws, may result in the production of mixed waste (radioactive and hazardous RCRA-regulated waste), which would increase disposal costs.

[10] Neutrons elastically scattering on hydrogen can lose more than half of their initial neutron kinetic energy. About 18 collisions with hydrogen are required to fully thermalize a 2-MeV neutron. Thus, hydrogen-containing materials are highly attenuating for neutrons.

# 6

# Comparison of Pulsed Fast Neutron Transmission Spectroscopy and FAA-Certified Explosives-Detection Systems

In this chapter, PFNTS is compared with the existing FAA-certified EDSs on the basis of their practicality for improving aviation security. All existing FAA-certified EDSs are based on x-ray CT, and all deployed certified EDSs are either InVision CTX-5000 or CTX-5000 SP systems. The data on PFNTS detection capability were provided to the panel by Tensor Technology (testing at the University of Kentucky) and the University of Oregon, and all of the PFNTS operational characteristics evaluated by the panel are based on Tensor Technology's MDNR design. The panel's intention, however, is not to compare Tensor Technology and InVision Technology but to make as comprehensive a comparison as possible between PFNTS and x-ray CT systems in terms of their acceptability for deployment in airports.

## PERFORMANCE

The motivation for deploying explosives-detection equipment is to improve aviation security. The selection of the equipment is determined by its ability to detect explosives and, at the same time, cause minimal disruptions of airline operations. The FAA's certification requirements reflect both of these requirements (FAA, 1992). Although the specifics of the FAA's certification requirements are classified, the factors that are measured are not. Perhaps the most important factor is the $P_d$ (probability of detection). To pass certification testing, an EDS must be able to detect various explosives configurations at a rate determined by the FAA. Because a high $P_{fa}$ (false alarm rate) could impede airline operations, the FAA has also set a requirement for a minimal $P_{fa}$. Finally, the FAA requires a throughput rate of 450 bags per hour.

The detection of explosives is the fundamental performance criterion for comparing competing technologies. The results of certification testing of the InVision CTX-5000 are classified, and some results from blind tests of the Tensor MDNR cannot be presented because they are subject to the provisions of 14 CFR 191.1.[1] Nevertheless, because the critical aspects of the test protocol for the MDNR blind tests were consistent with those used for certification testing of the CTX-5000 SP, the $P_d$ can be compared in a general sense. The blind test results that are available for the MDNR are shown in Table 3-2. The CTX-5000 has passed FAA certification testing and, therefore, meets the FAA's detection requirements. Although the MDNR has not been submitted for certification testing, the blind test results suggest that the $P_d$ for all categories and configurations of explosives required for certification, with the exception of Class A explosives, would be acceptable. The inability to detect Class A explosives is the most significant deficiency of the MDNR. Since the blind tests were conducted, Tensor has refined the detection algorithm on simulated data (Tensor Technology, 1998c); however, in the opinion of the panel, simulation results are not acceptable substitutes for actual test results.

Based on the results of laboratory blind tests of the MDNR (Tensor Technology, 1998b) and certification testing of the CTX-5000 SP (FAA, 1996a), there is no evidence that the MDNR significantly exceeds the detection performance of the CTX-5000 SP for any category of explosives. Furthermore, the cumulative[2] $P_d$ for the CTX-5000 SP is higher than for the MDNR. Although some field test results suggest that, in some cases, the detection performance of the CTX-5000 SP combined with an operator may be lower than the performance level of the MDNR, these data cannot be used for comparison because the MDNR has not undergone field tests.

[1] Information subject to the provisions of 14 CFR 191.1 is sensitive but not classified. Information determined by the FAA to be sensitive may not be released without the written permission of the Associate Administrator for Civil Aviation Security (ACS-1), Federal Aviation Administration, Washington, DC 20591.

[2] The cumulative $P_d$ is the $P_d$ of an EDS averaged over all explosives categories.

A low $P_{fa}$ is another important requirement for FAA certification. The CTX-5000 SP met the FAA's $P_{fa}$ requirement during certification testing. The $P_{fa}$ for the MDNR during blind testing at the University of Kentucky would not have passed certification testing, although the $P_{fa}$ of the PFNTS in blind tests at the University of Oregon was much lower than the requirement for certification. The bag contents used for the PFNTS blind tests were not the same as the bag contents used during certification testing of the CTX-5000 SP, however, which makes direct comparisons difficult to make. Furthermore, $P_{fa}$ and $P_d$ are closely correlated and should not be addressed in isolation. The panel believes a credible comparison would require complete receiver-operating characteristic curves for both systems. Unfortunately, this level of characterization is not available for either the CTX-5000 SP or the MDNR. It has been reported that the automated[3] $P_{fa}$ for the CTX-5000 SP is higher in the field than it was during certification testing and that the automated $P_{fa}$ of the MDNR during blind testing was lower than the automated $P_{fa}$ for some of the deployed CTX-5000 SP units (DOT, 1998).

Alarm resolution is critical to evaluating the performance of a system in an airport. In field tests of the CTX-5000 SP, the issue of alarm resolution was addressed by operator assistance and other airport-specific measures. The issue of alarm resolution for PFNTS has not been adequately addressed in existing studies. However, the $P_{fa}$ for the CTX-5000 SP-operator combination in field tests was lower than for the MDNR in laboratory tests. Because the spatial resolution of the PFNTS image is low, operator intervention does not lower the $P_{fa}$. Therefore, another method of alarm resolution would have to be found.

The last major performance metric is throughput rate. The CTX-5000 SP has a throughput rate of 225 bags per hour per machine; in the certified configuration of two CTX-5000 SP instruments, it passes the FAA's requirement for a combined throughput rate of more than 450 bags per hour. Tests of the MDNR concept at the University of Kentucky using a linear accelerator demonstrated a throughput rate of 16 bags per hour. The rate was severely limited by the linear accelerator, which had a low current and, therefore, low neutron fluence. Tensor estimates that with an Ebco TR9D cyclotron accelerator (which operates at a current roughly 30 times higher) the scan time could be as low as eight seconds per bag, which translates to a throughput rate of 450 bags per hour, which is consistent with the FAA requirement. In the panel's judgment, the MDNR could attain the throughput rate required for certification.

[3] The automated $P_{fa}$ is the false alarm rate prior to alarm resolution by an operator.

## OPERATIONS

It could be argued that the most significant operational consideration for deploying explosives-detection equipment is overall cost, which includes purchase price, installation costs, personnel costs, consumables costs, and maintenance costs. Ultimately, every operational characteristic affects cost. For example, if false alarm resolution is a slow, arduous process, then the throughput rate goes down, which could cause flight delays and, therefore, increased costs. Various operational aspects of PFNTS and CT systems are evaluated in the following sections.

### Costs

The first cost incurred by the government or the air carrier is the purchase price of the equipment. As shown in Table 6-1, an InVision CTX-5000 SP costs \$1 million; the projected purchase price of the initial MDNR is \$2.7 million.[4] The installation cost for the CTX-5000 SP ranges from \$20,000 for a lobby installation with no modifications to the baggage line to \$3 million for a complex installation fully integrated into the baggage-handling system, which requires extensive modifications to the baggage line. The average installation cost for "stand-alone" units was \$20,000 to \$150,000 per system. Tensor's estimates of the installation cost for the MDNR for the simplest possible airport baggage-line configuration is \$323,000 to install the vaulted area to house the MDNR and another \$175,000 to modify baggage lines (Tensor Technology, 1998a). Based on the experience of deploying the CTX-5000 SP, the panel is skeptical that the MDNR could be installed for the projected cost, especially if a separate facility is required to house the MDNR.

Tensor estimates it will cost \$39,000 a year for manpower to operate the MDNR, \$15,000 a year for consumable supplies, and \$6,000 a year for maintenance (Table 6-1). For deployed CTX-5000s, the manpower costs are \$150,000 a year,[5] and maintenance costs are \$48,000 to \$90,000 a year.

### Reliability and Maintenance

Redundancy is a factor that should be considered when comparing CTX-5000 SP installations with a proposed PFNTS system. In many airports, two or more CTX-5000 SP units have been installed, either to increase throughput or to inspect international transfer baggage. The benefits of having multiple CTX-5000 SP units are obvious. If mechanical problems, baggage jams, or other problems render one unit inoperative, the entire airport operation does not have to be shut down. The size and siting constraints of PFNTS systems

[4] In quantities of 10 or more, Tensor Technology projects that the MDNR could be sold for \$1.5 million.

[5] This cost includes a full complement of security screeners who may perform other duties when they are not operating the CTX-5000.

TABLE 6-1 Baseline Characteristics/Attributes Used in This Assessment

| Attribute | CTX-5000 or CTX-5000 SP | Comments | Projected Data for MDNR | Comments |
|---|---|---|---|---|
| Unit cost ($ million) | 1.0 | | 2.7 | Initial unit |
| | | | > 25 | Development cost<br>Tensor states $1.8M for assembly and testing of the MDNR prototype. (The additonal cost is the panel's estimate of the development cost before airline deployment of a PFNTS production version.) |
| | | | 1.5 | In quantities of 10 or more. |
| Installation cost ($ thousand) | 20 ± 10 | Lobby installation with no modifications. | 323 | Tensor's estimate for vaulted area. |
| | 150 ± 50 | Lobby/behind ticket counter installation with modifications. | 175 | Tensor's estimate to modify baggage lines. |
| | 2,000–4,000 | Fully integrated installation including cost of one CTX-5000 SP. | | |
| Complexity of installation | Easy to difficult | CTX-5000 series EDSs have been installed in lobbies, behind ticket counters, and in baggage lines. (One baggage-line installation was in a mezzanine [2nd floor].) | Difficult | Deployment of the MDNR (except on ground level) is probably not feasible because of weight considerations. |
| Compatibility with baggage-handling system | Yes | Installations in airport terminal lobbies and behind ticket counters are relatively straightforward. Installations that require integrating the CTX-5000 into baggage lines are more difficult. | Difficult | Placement in baggage-handling station or in separate building. Integration appears to be straightforward, but sharp bends are not compatible with bag movement, while straight paths are not compatible with radiation shielding. |
| $P_d$ (%) | Classified | | 85 | $P_d$ for Class B explosives was 94.4; for Class A explosives it was 51.8. |
| $P_{fa}$ (%) | Classified | | 25 | In Tensor blind testing. |
| Operations cost ($ thousand/shift/year), including overhead | 48–90 | Maintenance policy. | 90 | Manpower (Tensor states $39,000). |
| | 150 | Manpower | 15 | Consumable supplies. |
| | 10 | Operator training. | 6 | Maintenance |
| Operators (number/machine/shift) | 1 | | 1 | Cyclotron operator. |
| Ancillary manpower | 0 | Fully integrated installation. | | |
| | 1 | Lobby installation and partially integrated installation. | | |
| | 2 | Behind counter; one bag handler on front and one on back end. | | |
| Educational requirements (years) | None | No educational requirements, but InVision's recommends that operators speak English and not be colorblind. | 2 | Associate's degree (or equivalent) in a technical field. |

TABLE 6-1 Continued

| Attribute | CTX-5000 or CTX-5000 SP | Comments | Projected Data for MDNR | Comments |
|---|---|---|---|---|
| On-the-job training requirements | 1–2 weeks | 1 week for foreign installation. 2 weeks for domestic installations. FAA requirement sometimes stretches up to 4 weeks of training. | 6–12 weeks | Specialized training. Tensor suggests 6 weeks. |
| Bag throughput (automated bags/hr) | 200–250 | CTX-5000 SP (111 bags/hour highest throughput, including alarm resolution, observed in the field). | > 200 | Tensor estimates potential of 1,674 bags/hr. |
| | 270 | CTX-5500 DS SURE software. | | |
| | 380 | CTX-5500 DS CERT software Both softwares certified by FAA. | | |
| Down time (%) | 2 | InVision prefers to describe as “up-time” (98%). | < 2 | Tensor estimate supported by Ebco Technology (manufacturer of the cyclotron in the proposed Tensor MDNR design) does not include down time of support equipment. |
| Film safe | No | For ASA > 400 damage to film is evident in prints. | | |
| Film tolerant | Yes | 100, 200, and 400 ASA damage in negatives but not in prints. Appears to be agreeable to vast majority of travelers. | unknown | Tensor states yes, but more testing required. |
| Alarm resolution | Relatively easy | CTX-5000 SP provides a cross-sectional image useful for resolving alarms. | Moderate to difficult | Image produced by stand-alone MDNR is not useful for alarm resolution. Bag could be imaged with x-rays to resolve alarms. |
| System weight (tonnes/tons) | 4.3 (4.7) | | 657 (723) | Tensor estimate. |
| System footprint ($m^2/ft^2$) | 11.3 (125) | Machine footprint is 5 $m^2$ (56 $ft^2$) (length = 4.45 m [14.7 ft] and width = 1.89 m [6.25 ft]) with an additional 6.2 $m^2$ (69 $ft^2$) required for ramps and console. This does not include space required for baggage handling or the INVIROPAK. | 93 (1034) | Tensor estimate of footprint (length = 11.8 m [39 ft] and width = 7.88 m [26 ft]). Does not include some support equipment requiring about 3.6 $m^2$ (40 $ft^2$). |
| System height (m/ft) | 2 (6.7) | | 3.5 (11.5) | |
| Power requirements | 12 kVA; 50–60 Hz; 350–510 V 3 phase | | 77 kW | Tensor estimate. |
| Temperature requirements | 10–40 °C (50–104 °F) | | 18–25 °C (64–77 °F) | Temperature limits from Ebco literature. Higher temperatures may be permissible. |
| Relative humidity requirements (%) | < 80 | | < 70 | |

TABLE 6-1 Continued

| Attribute | CTX-5000 or CTX-5000 SP | Comments | Projected Data for MDNR | Comments |
|---|---|---|---|---|
| Utilities | Electricity and telephone connections. | | Not clearly indicated in conceptual design. | Clean air, temperature-controlled water, stable power with short cable lengths from supply. Tensor states no problem. Certainly practical but may require additional support equipment located at MDNR. |
| Mean time between failure (hours) | 722 | Determined during certification testing. Value not obtained from field experience. | Unknown | Estimate that filament in cyclotron must be changed once per month (which takes approximately 15 minutes). Operational data lacking. |
| Mean time to repair (minutes) | 54 | Determined during certification testing. Value not obtained from field experience. | Unknown | |
| Safety concerns | Minimal to none. | Meets FDA cabinet x-ray machine specifications. | Severe | Radiation safety. Shielding is very practical, but safeguards are required to ensure very high radiation levels within cyclotron shielding and that apertures are monitored. Requirements are similar to hospital requirements for cyclotron production of radioisotopes. |
| Radiation monitoring requirements | None | Meets FDA cabinet x-ray machine specifications. | Yes | TLDs for monitoring personnel are required. Machine survey is required. Area survey is required. |
| U.S. Nuclear Regulatory Commission license required | No | | No | |
| Radiation shielding issues | None | Unit is self-shielded. | Important | Additional studies required to validate conceptual design. Shielding appears to be feasible with minor changes in design. |
| Public perception of health risks | Low | Similar to current x-ray systems. Main public concern is that it is not film safe. | Moderate | Neutron irradiation of bag will generate some protest, but safety can be assured. System does not interface directly with passengers. |

may prohibit the installation of more than one unit at a single airport. Therefore, if the unit were out of service—for any reason—the baggage-screening operation would be severely compromised.

The deployed CTX-5000 SP has 98 percent up time (i.e., 2 percent down time). The mean time between failure (in laboratory testing) for the CTX-5000 is 722 hours, and the mean time to repair is 54 minutes (FAA, 1996a). Tensor estimates that the cyclotron would require ~2 percent down time for maintenance: the filament in the cyclotron would have to be replaced once a month (i.e., mean time between failure is 720 hours), which takes about 15 minutes. The mean time to repair is typically 8 hours for cyclotrons similar to the ones in the Tensor design. Ebco Technologies, the manufacturer of the cyclotron accelerator in the Tensor MDNR design, agrees with Tensor's estimates. The estimates do not address down time associated with the detector array or support equipment.

## Alarm Resolution

The Tensor report does not address the issue of alarm resolution, a very important consideration in the selection and fielding of explosives-detection equipment. The MDNR is at a significant disadvantage in this respect compared with existing image-based detection systems, such as the CTX-5000 SP. The coarse grid image provided by a PFNTS-based system (see Figures 3-1 and 3-2) was of little help to an operator for resolving an alarm during blind tests; thus, the automated $P_{fa}$ was nearly identical to the $P_{fa}$ with an operator present. The CTX-5000 is of more help for alarm resolution because it produces a sharper image for the operator to view. The utility of this image for alarm resolution has been confirmed by the significantly lower $P_{fa}$ of the operator–CTX-5000 combination than the automated $P_{fa}$ of the CTX-5000 alone (FAA, 1997a).

Even if the MDNR's $P_{fa}$ is similar to the rates of certified EDSs, the alarm resolution issue strongly suggests that existing CT technology is preferable. To address this issue, the $P_{fa}$ of the MDNR would have to be much lower than the rate required by the current certification specifications. Even if the MDNR $P_{fa}$ could be lowered to 1 percent, which is beyond reasonable expectations based on the existing test results, an MDNR in an airport would have to be considered as just one level in a multilevel explosives-detection system. Although additional explosives-detection equipment could be located at a different place along the baggage line, the poor spatial resolution of the MDNR technology would probably require that a higher resolution technique be located in close proximity to the MDNR to resolve alarms. In other words, the MDNR would probably have to be integrated with a high-resolution imaging technology, such as an advanced x-ray or CT x-ray system. The complementary aspects of the x-ray and the PFNTS detection approaches suggest that this combination could be very effective.

## AIRPORT INTEGRATION

The location of security equipment on airport property is a primary concern of air carriers and airport operators. Site surveys to identify space for the CTX-5000 SP presented many challenges to the FAA's SEIPT, the airlines, and airport operators. Airport space is at a very high premium, especially at busy airports where security equipment is needed most. Finding the 945 $m^2$ (9,360 $ft^2$) of floor space required for a PFNTS would be much more difficult than finding the 6.7 $m^2$ (67 $ft^2$) required for the CTX-5000 SP. Another constraint would be that only the ground floor could support the 657-tonne (723-ton) weight of an MDNR installation without major floor-support construction. Thus, an MDNR could not be installed in most, if not all, ticket/check-in areas of an airport terminal building, which has been shown to be the most desirable location for screening baggage in certain airports. Another concern is that airlines and airport tenants regularly exchange space and modify terminal configurations. The deployment of a PFNTS-based explosives-detection device would inhibit this flexibility because of the extreme weight and size of the equipment.

### Baggage Screening

The complex process of handling baggage in U.S. airports depends on cooperation between air carriers, including the sharing of baggage-handling systems, which are very labor intensive. In general, the only automated portions of most baggage-handling systems are baggage conveyor belt(s), which run from ticket counters or curbside check-ins to the baggage sorter location, and automated on-demand baggage tag printing. Even "automated" baggage conveyor systems, however, require manual intervention because of baggage jams, mechanical failures, oversized bags, and special articles (e.g., animal kennels). The integration of explosives-detection equipment into the baggage-handling system adds considerable complexity to the operation.

The MDNR proposed by Tensor Technology, which produces neutron and gamma radiation, would require that the device be enclosed in a 528-tonne (581-ton) concrete vault with radiation locks and a heavily shielded door (Tensor Technology, 1998a). In the preliminary design, baggage is shown flowing through the unit on an automated belt system with six 90-degree turns inside the vaulted enclosure (see Figure 5-2). A mechanical device would place each bag in the proper (upright) position for scanning. The bag would be stopped, scanned, and then restarted to exit the system. By today's standards, this is a "complex" baggage-handling system. By contrast, the InVision CTX-5000 SP has been successfully installed in 18 U.S. airports in lobby installations, partially integrated installations (e.g., behind the ticket counter), and fully integrated installations (e.g., integrated into the baggage line). Although all of these installations encountered problems, baggage-handling operations were essentially uninhibited by the CTX-5000 SP (NRC, in progress).

The performance of existing automated baggage systems—without the added complexity of explosives-detection equipment—clearly shows that bags occasionally get "jammed" together, requiring that the belt be stopped and that manual (human) intervention be used to clear the jam. Experience has shown that the more complex the system is, the more likely jams and mechanical problems are to occur. Therefore, it is likely that baggage jams would occur inside the vault enclosure in the proposed MDNR design, especially because of the changes in direction in the baggage line. Health and safety guidelines would prevent an airline employee from immediately entering the MDNR enclosure to clear a jam. By the same reasoning (and by experience), because the design of the CTX-5000 SP involves no changes in the direction of the bag line inside the CTX-5000 SP structure, baggage jams are unlikely to occur. Furthermore, an operator can easily clear a baggage jam at the entrance or exit of the CTX-5000 SP.

Redesigning the baggage-handling component of the MDNR could reduce, but not eliminate, the likelihood of baggage jams inside the MDNR enclosure. Even a redesigned system could not eliminate the delay in getting an employee safely into the enclosure to clear a baggage jam or correct a minor mechanical problem. This delay would have a negative impact on system operating performance, especially on throughput.

### Cargo Screening

The 1996 Valujet accident brought safety and security issues related to cargo shipments on passenger aircraft to national attention (DOT, 1998). Passenger aircraft carry nearly 60 percent of all air cargo,[6] and the air cargo industry in the United States involves linkages between 4,000 air carriers, 3,000 forwarders of air freight, 4,000 repair stations, and 70,000 shippers of dangerous goods (DOT, 1998). Although these issues were reviewed by the White House Commission on Aviation Safety and Security (1997) and by Congress, policies and regulations regarding security screening of cargo and mail are still in a state of flux and will not be reviewed in this report. Based on information submitted to the panel by the FAA, however, the panel assumed that some form of cargo screening for explosives will be required in the future (Fainberg, 1998).

As the Tensor report points out, the air cargo problem is difficult to address because of the large variety of sizes and substances that are shipped (Tensor Technology, 1998a). Configurations of cargo, which vary by type of aircraft and cargo shipper, include containerized cargo (e.g., LD-3 containers), palletized cargo, and loose-loaded cargo (sometimes called bulk-loaded cargo). Furthermore, cargo make-up may be done either at the airport or at the cargo shipper's facility, which has implications for the explosives-detection equipment that can be used.

Tensor's assessment of the applicability of the MDNR for cargo inspection is based on the assumption that cargo is shipped in LD-3 containers and that each container contains a single class of material (e.g., books or frozen fish). This assumption is based on a rough characterization of air cargo provided by the FAA (1996b), and on this basis the analysis in the Tensor report is valid. However, in actual practice cargo may be shipped in other types of containers, such as the LD-2 container, or as palletized or loose cargo. To complicate matters further, the cargo in a single container may not be uniform. Containers could carry a variety of cargo items. Tensor's assessment, which is based on a few experimental tests of the MDNR with small packages and theoretical extrapolations, may not be valid for all of these variables. In fact, the MDNR has not been tested for full LD-3 containers or other containers of comparable size. Therefore, the capacity of PFNTS to screen cargo for explosives cannot be assessed until additional tests have been conducted.

As presently configured, the CTX-5000 SP cannot scan an LD-3 container for explosives because the opening of the CTX-5000 SP is too small to fit an LD-3 container inside. Even if a CT-based explosives-detection system capable of scanning an LD-3 container were developed, it is highly unlikely that it would be effective for detecting explosives because the x-ray attenuation in a container the size of an LD-3 would reduce penetration to the point that an analysis of the contents would not be feasible. When mail, packages, and other cargo are placed on pallets or in containers at the airport, items could be screened by the CTX-5000 SP prior to make-up. Although the CTX-5000 SP may be as effective in detecting explosives in suitcase-sized (or smaller) cargo as it is for passenger baggage, it has not been validated for this use (it should be noted that the panel did not have any data on the effectiveness of the CTX-5000 SP for detecting explosives in cargo).

Many of the issues for integrating either the CTX-5000 SP or the MDNR into airports for cargo screening are the same as for passenger bag screening. The restriction on installing the MDNR in lobbies would be less relevant, however, because cargo would be screened elsewhere in the airport. Size, weight, cost, and performance, as well as the impact on airline operations, would all be important considerations in selecting a technology for cargo screening.

Based on available data, the panel believes that the deployment of the MDNR or any other PFNTS-based equipment for cargo screening cannot be justified because only a fraction of the total cargo loaded onto an aircraft would be effectively screened. Thus the cost-benefit ratio would be high. The panel also concluded that there is no conclusive evidence that the CTX-5000 SP is an appropriate system for screening cargo.

## LICENSING AND REGULATIONS

The licensing and operating regulations for x-ray-based EDSs are well understood and are routinely implemented in airports. The federal regulations for the generic operation of a PFNTS-based system with an accelerator and a neutron source are documented in regulatory acts and in the Code of Federal Regulations (CFR): 10 CFR 834, Radiation Protection of the Public and Environment; and 10 CFR 835, Occupational Radiation Protection. Regulations that address the safety of personnel and protection of the environment include the Clear Air Act, the Clean Water Act, the Safe Drinking Water Act, the Radiation Control for Health and Safety Act of 1968, the Occupational Safety and Health Act of 1970, the National Environmental Policy Act, and RCRA.

The CFR addresses the implementation details for the production of medical isotopes and gamma irradiators. However, because the radiation-producing system of a PFNTS would not be used for medical purposes and does not have a sealed gamma source, it is not covered by these regulations. Because the use of accelerators for explosives detection is new, no regulatory implementation details have been developed. In the absence of regulations governing the use of neutron-producing accelerators for explosives detection, the panel used the regulations for gamma irradiators as indicators of possible requirements. The scope and potential

---

[6] Air cargo can include the following items: airmail (e.g., sacks, flats, and boxes); express packages; COMAT (company materials); COMAIL (company mail); diplomatic pouches; courier pouches; and baggage (shipped as cargo).

impact of some of these regulations are outlined in Box 6-1. If the laboratory-based PFNTS technology were moved into a public area, such as an airport, there would probably be changes in the current regulatory requirements and, perhaps, more specific implementation guidelines, as there are for gamma irradiators used to sterilize materials. In addition to federal regulations on the treatment and control of radioactive materials and sources, some states also have licensing and registration requirements for accelerators.

The details of the system design and operation of the MDNR are at the level of conceptual design but do not address regulatory operational requirements and cannot be used as a basis for evaluating the impact of regulations. Until detailed documentation has been prepared on an engineering-level design, the compliance of PFNTS designs with the regulatory requirements must be considered a significant, but not insurmountable, impediment to acceptance by airlines and airports.

Regulatory hurdles for PFNTS will begin with the design and construction of the system and extend to the operational

## BOX 6-1
## Selected CFR Regulations Relevant to PFNTS

CFR 36.29 specifies that irradiators with automatic product conveyor systems must have a radiation monitor with an auditable alarm to detect loose radioactive sources that are carried toward the product exit.

*This regulation applies to irradiators that contain sealed gamma-emitting sources. Moderate-energy gammas (< 5 MeV) cannot activate most materials. The radiation monitor for irradiators is focused on early detection of possible leaks from the sealed gamma source. Given the cautionary nature of these regulations (the sealed gamma source material must be doubly encapsulated), a prudent design of a neutron-producing PFNTS system capable of activating material in the vault area will almost certainly incorporate an exit detector to inspect all outgoing bags. An exit monitor would be a prudent measure even if calculations suggest that the normal bag neutron illumination would not reach regulatory limits for activated material because activated room material may have fallen onto the bag. This conservative design is consistent with the design of the thermal neutron analysis system, which also has an exit detector.*

CFR 36.51 lists detailed requirements for training operators of irradiators. Regulations call for instruction on applicable radiation safety issues, a written test, on-the-job or simulator training, and annual safety reviews.

CFR 36.53 requires written procedures for most activities, including the radiation surveys when entering or leaving irradiator radiation rooms, irradiator operations, and facility inspections.

CFR 36.55 requires radiation monitoring of personnel with film badges or thermoluminescent dosimeters (TLD) at irradiator facilities. The TLD processor must be accredited by the National Voluntary Laboratory Accreditation Program. TLDs must be processed at least quarterly and film badges at least monthly.

CFR 36.57 requires radiation surveys of the area outside the irradiator shielded area at intervals not to exceed three years or whenever the source strength has been changed. This would imply shielding surveys on recently installed cyclotrons and every three years of operation as long as the maximum permissible deuteron beam current is not increased.

CFR 36.57 requires portable survey meters at irradiators to be calibrated at least annually.

CFR 36.63 requires "an irradiator operator and at least one other (trained) individual" to be present on site. This suggests that a PFNTS system might require that two trained persons be on site during operations. Cross training of operations personnel could ensure that the second person could perform other duties and not just be on call. A radiation safety officer might also be required to be on call.

CFR 36.81 lists detailed requirements for keeping records and requirements for accident reports for irradiator facilities, including requirements for records related to decommissioning.

environment. There are no regulations that appear to require an Environmental Impact Statement (EIS) for the MDNR. However, based on experience with the thermal neutron analysis system, completing an EIS for the initial MDNR units would be prudent. Because public perception will be an important consideration in the decision to deploy a PFNTS system, the EIS would assure the public that the system has been designed and will be operated safely and in compliance with all applicable regulations.

Some states have regulatory requirements for radiation-producing equipment (e.g., accelerators) that must also be addressed. Although the panel has no reason to believe that an MDNR would be incompatible with the regulations in any state, the differences among the regulations could be an impediment to the acceptance of the MDNR at an airport. Once an engineering-level design becomes available and implementation details have been formulated to ensure compliance with federal regulations, state regulatory agencies would have to be consulted to determine if design changes would be required for compliance with the full range of state regulations. Although none of the anticipated federal or state regulatory requirements is clearly incompatible with the efficient airport implementation of PFNTS technology, if establishing the infrastructure to ensure compliance with regulatory requirements were difficult and costly it could discourage airline and airport authorities from selecting PFNTS explosives-detection technology (even if the system met the FAA certification criteria).

# 7

# Conclusions and Recommendations

At the current stage of development, constructing an airport-compatible prototype of a PFNTS-based explosives-detection device, such as the MDNR, would be ill advised. Current PFNTS designs appear to be at a substantial disadvantage compared to x-ray CT-based technologies that have already been certified. Based on the information currently available, an installed PFNTS-based device would cost considerably more than currently available FAA-certified EDSs. Even if the PFNTS per-unit costs could be brought down to a level comparable to those of CT-based systems, the PFNTS system would still be at a disadvantage because it requires much more space to install and is considerably heavier, which would make it much more difficult to install in existing airports; in fact, the entire baggage transfer system would have to be redesigned and reconstructed.

The promise of PFNTS—and other element-specific technologies—is their potential to detect a wider variety of threats and lower threat masses than are currently included in the FAA's EDS certification standards. Thus, a critical remaining step in the development of PFNTS technology is characterizing the range of potential threats and threat masses it could reliably detect. This kind of research would not be practical with a prototype system designed to meet current EDS performance certification standards and packaged in a compact configuration for integration into an airport environment.

Another attribute of PFNTS is its capability of responding to changing explosive threats to airline security. It seems reasonable to expect that eventually the certification standards will be tightened to require detection of smaller explosive quantities and a wider range of explosive types with less defined densities. As terrorists become more sophisticated in the placement of explosive devices, the threat quantities of interest could become smaller. The original categories for EDS certification testing reflect the explosives of choice by terrorists at the time the certification standard was developed. The promise of PFNTS is not that it can detect explosives at the current certification standards but that it has the "potential" to detect smaller threat quantities and a wider range of explosives. Therefore, research should be focused on characterizing the potential performance of PFNTS rather than on optimizing the airport-installed performance of a current point design. This type of research could best be done in a research setting rather than in an airport environment with a prototype designed to meet current certification standards.

## QUESTIONS POSED BY THE FAA

At the panel's first meeting, the FAA asked that four questions be addressed during the course of this study. The panel and the NRC staff determined that these questions fit within the Statement of Task.

**Question 1.** Given the choice, would airlines select the PFNTS instead of equipment based on currently available x-ray CT for checked baggage inspection?

**Answer.** No, the airlines would not choose a PFNTS-based explosives-detection device for three reasons. First, PFNTS has not demonstrated an ability to meet the FAA's certification requirements for detecting Class A explosives. Second, the area detector configuration has not demonstrated an ability to meet the FAA's false alarm requirements. Third, because of integration difficulties that would be caused by the size, weight, and safety precautions attendant upon the deployment of a PFNTS-based device, the airlines would choose currently available x-ray CT-based EDSs.

**Question 2.** If so, is their preference for this technology strong enough to justify the remaining costs to develop this technology (estimated to be \$20 million to \$30 million)?

**Answer.** Not applicable.

**Question 3.** Does PFNTS have any realistic potential for application to full cargo container inspections?

**Answer.** No. Based on Tensor's analysis of PFNTS for cargo screening of LD-3 containers containing single items (a reasonable projection given the sparseness of existing test data), PFNTS could only interrogate 57 percent of air cargo shipped in LD-3 containers. The other 43 percent contains large amounts of hydrogenous and, therefore, highly neutron-attenuating material, rendering PFNTS ineffective for screening. Further complicating the use of PFNTS for cargo screening is that containerized cargo is not always uniform in composition. The likelihood of false alarms from nonhomogeneous containerized cargo assembled by cargo consolidators has not been determined. Finally, the capability of PFNTS to detect explosives concealed in thick containers (e.g., the LD-3) has not been experimentally verified. Current estimates on neutron attenuation (based on crude exponential algorithms) are sufficiently accurate to raise concerns about highly attenuating hydrogenous cargo. These estimates do not treat beam divergence from neutron scattering in the cargo and are, therefore, not sufficient to validate the PFNTS detection for low-attenuating scenarios.

**Question 4.** What experiments, if any, should be pursued in the near future to further define this potential?

**Answer.** Because PFNTS does not appear to be either practical or currently desirable for airport deployment, the panel does not recommend that experiments addressing the airport integration of PFNTS be pursued at this time. However, experimental verification of Tensor's simulation of PFNTS performance for cargo screening might be useful.

## PROTOTYPE

A critical limitation of PFNTS-based explosives-detection technologies is their inability to detect Class A explosives at the $P_d$ level required for FAA certification. Other limitations are the high $P_{fa}$ demonstrated for area detectors and their unproven capacity for resolving alarms. Unless and until these limitations are overcome, there is no reason for the FAA to address the technical or operational issues associated with integrating the technology into an airport setting. Even for the current certification testing requirements for Class B explosives (as opposed to Class A explosives), PFNTS-based technologies would not be selected by airports for primary screening of carry-on baggage or checked baggage because of the difficulties of integrating these technologies into existing baggage lines and the operational issues associated with ensuring the safe operation of radiation-producing accelerators that meet regulatory requirements.

To date, no one has demonstrated that any PFNTS-based technology, including the MDNR design proposed by Tensor Technology, could be feasibly integrated into existing airport operations. The 7.9 m × 12.6 m (26 ft × 42 ft) rectangular space requirement described in the MDNR design could not be provided easily at most airports. Furthermore, few if any existing airport terminal structures could accommodate the 657-tonne (723-ton) weight without significant alterations. The size and weight of PFNTS-based explosives-detection devices are largely dictated by shielding requirements and by the necessary distance between the accelerator and the detector array for time-of-flight energy resolution. The panel believes it is unlikely that a PFNTS-based device could be designed with a much smaller footprint than the MDNR proposed by Tensor Technology. Therefore, x-ray CT-based systems are a better choice for detecting current explosive threat levels.

If the PFNTS system were configured for the smallest possible size, a complex baggage belt transfer system would be required to move baggage through the unit and to orient it properly in front of the detector array. In actual operation, this complexity is likely to lead to time-consuming baggage jams that would have to be cleared manually. The time required to shut down the system and enable an operator to safely enter the shielded environment and clear a jam would substantially reduce the throughput of the system.

Even though the panel believes a PFNTS-based device could be operated safely, the public, and perhaps even airline employees, might believe the system posed health risks. Because of the perceived risks and the physical size and weight of PFNTS-based devices, operations personnel could decide to locate the device in a remote facility, either a stand-alone facility or a separate building outside of the terminal complex, which would inhibit passenger involvement in resolving false alarms. Therefore, a PFNTS system would have to demonstrate a much lower $P_{fa}$ than the current certification standard to offset the difficulty of resolving alarms.

The potential of PFNTS for screening small loose cargo packages has not been sufficiently explored using existing research accelerators, and the existing characterization of air cargo (type, size, weight, delivery constraints, method of delivery to the airport) is not sufficient to develop a valid testing protocol. Furthermore, although the FAA may have intelligence data that support determining the type and mass of an explosive threat, the agency has not endorsed a specific threat definition that could be used as a basis for evaluating the potential of any explosives-detection technology for cargo inspection. Based on cargo characterization data and analytic estimates of neutron attenuation, the panel concluded that PFNTS does not have a realistic potential for screening the full spectrum of cargo containers and pallets.

**Recommendation.** The FAA should not fund the development of a prototype of a multidimensional nuclear radiometer-based explosives-detection device.

**Recommendation.** Based on the current explosive threat levels and the state of the art of pulsed fast neutron transmission spectroscopy (PFNTS) technology, the FAA should not deploy PFNTS technologies or devices for primary screening of carry-on baggage, checked baggage, or cargo.

**Recommendation.** At this stage, the FAA should not fund the development of an airport test facility.

## RESEARCH

After reviewing the laboratory-demonstrated detection performance of PFNTS (Chapter 3) and comparing PFNTS-based devices to x-ray CT-based EDSs (Chapter 6), the panel concluded that deploying a PFNTS-based system would not be practical at this time. Indeed, laboratory testing has not demonstrated PFNTS to be technically desirable compared to the FAA-certified InVision CTX-5000 SP, CTX-5500 DS, and L3 Communications 3DX-6000, which are much easier to integrate into the airport environment. However, because the threat to aviation security is dynamic, the requirements for explosives-detection system certification may very well change over time; and, at some point, new explosives-detection approaches will probably be necessary. Among the technologies currently in development, PFNTS has shown the most promise of meeting more stringent certification testing requirements because of its element-specific detection technology. In other words, because PFNTS can determine multiple element-specific metrics of the materials it characterizes, it has the potential to detect lower threat amounts with a lower $P_{fa}$ and to adapt to new threats. Therefore, although the deployment of a PFNTS prototype (e.g., MDNR) is not desirable at this time, valuable research could be conducted on the application of PFNTS for explosives detection.

### Priorities

Existing neutron sources cannot readily provide the beam characteristics (high-current, low-radiation background) for significantly improving the quality of the test data. If innovative approaches to the use of currently available research accelerators cannot resolve this deficiency, research on the PFNTS technology for explosives detection may have reached a logical stopping point. If so, the experimental data (including baggage scan data from blind tests and complete documentation on the application of the detection algorithm to these scans), details of the detection algorithm, and suggested paths for refining the detection algorithm should be archived to preserve the knowledge gained from FAA-funded research. The technology could then be reevaluated if new security threats evolve for which current CT-based systems are inadequate and for which element-specific approaches might have the potential to meet aviation security requirements.

If funding is available for further research on PFNTS technology for explosives detection, the greatest benefit would be derived by addressing the current shortfalls of the technology rather than by developing and assembling a prototype MDNR unit that could be integrated into an airport. Because this panel has not studied and is not acquainted with the whole spectrum of research requests for explosives-detection technologies submitted to the FAA, the panel neither supports nor opposes the allocation of funds for further research on PFNTS. The recommendations in the following paragraph are based on the assumption that funds are available but should not be interpreted as endorsing such allocations.

The research recommendations in this report are directed at improving the detection performance of PFNTS-based explosives-detection technologies and do not address the practical limitations of deploying PFNTS-based devices. Because of the large mass of the cyclotron magnets and the extensive radiation shielding required for PFNTS, the size and mass of a PFNTS-based explosives-detection device can probably not be significantly reduced. These practical limitations should be considered in evaluating the potential of PFNTS for explosives detection in airports.

**Recommendation.** The research priorities for pulsed fast neutron transmission spectroscopy should be directly related to the current shortfalls of the technology. Three major problem areas that should be addressed are listed below in order of importance:

1. detection of Class A explosives
2. reduction of false alarm rates
3. development of alarm resolution procedures

The first priority for research on PFNTS is the development of a methodology for detecting Class A explosives, the one category of explosives for which PFNTS would not pass certification testing.

**Recommendation.** The pulsed fast neutron transmission spectroscopy detection procedure should be refined to detect Class A explosives reliably. These refinements will certainly involve using smaller target pixel sizes and may require multiple views or tomographic representations.

Reducing the false alarm rate of PFNTS will be critical for it to become a viable explosives-detection technology. Based on laboratory test data, the false alarm rate for a two-dimensional open detector array is uncertain. A strict interpretation of the existing laboratory blind test data, however, suggests that PFNTS would not pass the false alarm rate requirement for an FAA-certified EDS.

**Recommendation.** The FAA should validate the potential for a pulsed fast neutron transmission spectroscopy-based explosives-detection device with low false alarm rate and a high probability of detection. This will require resolving the sensitivity of the probability of false alarm to the detection algorithm (neural net versus B-matrix), evaluating the use of an area rather than a linear detector array, and assessing the effect of higher background neutron/gamma radiation on the detection algorithm.

Unless PFNTS is developed to the point that potential users have enough confidence in the system to shut down a terminal and call in a bomb squad for every unresolved alarm, alarm resolution will remain a critical issue. Because the image produced by transmitted neutrons is not sufficient for resolving false alarms, PFNTS will probably have to be paired with a high-resolution x-ray-based imaging technology for the purpose of alarm resolution.

**Recommendation.** The FAA should develop and validate a procedure for resolving false alarms for the pulsed fast neutron transmission spectroscopy technology that meets or exceeds the capabilities of deployed certified explosives-detection systems. This may involve pairing a pulsed fast neutron transmission spectroscopy-based device and an advanced x-ray-based high-resolution imaging system.

Several other issues should also be evaluated. The performance bounds for the PFNTS technology should be explored to determine the lowest amount of explosive that can be detected reliably while maintaining a low $P_{fa}$. In addition, the flexibility of the PFNTS approach should be validated experimentally by testing with a range of new threat materials. The applicability of PFNTS for screening cargo or small containers within cargo should be assessed. The research priorities to address these issues are listed below:

- Test the ability of the PFNTS algorithm to detect lower threat quantities and characterize the receiver-operating-characteristic curve for various threat quantities of existing classes of explosives.
- Characterize the detection performance ($P_d$ and $P_{fa}$) of PFNTS for small boxes in containerized and palletized air cargo.
- Quantify the influence of deuteron current stability, deuteron energy, and room size on detection performance.
- Quantify the performance of PFNTS technology for bags that have produced false alarms on existing certified EDSs, and characterize the performance for bags that contain explosives that were not detected by existing certified EDSs.
- Validate the flexibility and adaptability of the PFNTS detection algorithm to incorporate changes that address new threat agents, and characterize the time required to develop the expanded detection algorithm, highlight algorithm implementation problems if the new algorithm were implemented in older systems, and characterize any change in the $P_{fa}$ associated with the increased range of threat agents.
- Develop and demonstrate an automated hardware system that could orient general commercial airline bags so the PFTNS detection algorithm would be effective against Class A explosives (e.g., multiple views or standing rectangular bags on a side for the shortest cross section to facilitate neutron transmission).

## Facilities and Equipment

Consistent with the recommendations in the second interim report of the CCAS (1997), this panel believes that if the FAA continues to pursue research and development on PFNTS, current neutron sources, detector technology, and radiation modeling capabilities should be used. However, it appears that no existing research accelerator facility can meet the requirements for demonstrating the potential of PFNTS for explosives detection. These requirements include a 1-ns pulsed deuteron beam, high beam current (~50 μ-amps), a large experimental area consistent with a small bag pixel area and a long neutron time of flight path, and a low scattered neutron and gamma background that would provide PFNTS data with a high signal-to-noise ratio. If a research facility is acquired, it should not entail the development of a new accelerator; a commercially available accelerator should be acquired instead.

If a new accelerator is acquired, it should be configured to support a broad range of research activities. Rather than developing an airport-integrated system, further development of PFNTS for explosives detection should be focused on characterizing the potential of PFNTS technology and the sensitivity of the detection process to deuteron beam current stability, the selection of the deuteron energy, and the influence of the room size on the detector signal-to-noise ratio. The utility of an accelerator facility would be greatly enhanced if its design features could accommodate the research requirements of other postulated neutron-based applications.

Because the projected PFNTS research will probably be completed in the next few years, and given the expense of developing and constructing a new accelerator-based neutron test facility, the FAA should look for partners (e.g., Defense Advanced Research Projects Agency, U.S. Customs Agency, National Institutes of Health, or the National Science Foundation) to ensure the long-term usefulness of the facility. Locating the facility in a university environment might be the best way to promote broad cooperation and multidisciplinary research and ensure its long-term utility. In this way the facility would contribute to the education of students and promote low-cost innovative experiments that might lead to new uses of neutrons for explosives detection, as well as other applications. The panel does not support or oppose the allocation of funds for a research facility. All recommendations that require government funding are offered on the assumption that funds are available and should not be interpreted as recommendations in support of such allocations.

**Recommendation.** If the FAA acquires an accelerator to meet the testing requirements of PFNTS, it should be configured to support a broad range of research activities.

**Recommendation.** To ensure the availability of an accelerator research facility to a diverse range of researchers, the facility should be located in a university environment and not at an airport or private industrial site.

**Recommendation.** To ensure the flexibility of an accelerator research facility to address near-term and far-term research, it should have the following attributes:

- the capability of generating monoenergetic neutrons from the $^{2}H(^{2}H,n)^{3}He$ (DD) reaction in addition to the broad-energy "white" neutron spectrum from the $^{9}Be(d,n)^{10}B$ reaction required by PFNTS
- a large experiment room with configurable shielding walls to permit changing the size of the room and varying the scattered neutron background
- shielding in the experimental room and an approved experimental envelope consistent with the operation of a small 150-keV $^{2}H(^{3}H,n)^{4}He$ (DT) sealed-tube source in conjunction with other explosives-detection methods
- compatibility with other missions, such as the production of short half-life radioisotopes that can be used for research on the production and use of medical radioisotopes, which may entail accelerating protons as well as deuterons
- staff and faculty who can function as a core group for research activities and a radiation metrology laboratory to support characterization of the neutron field

# References


Chmelik, M.S., R.J. Rasmussen, R.M.S. Schofield, G.E. Gieger, H.W. Lefevre, J.C. Overley, and C.J. Bell. 1997. Analysis of Blind Tests for Explosives in Luggage through Fast Neutron Transmission Spectroscopy (FNTS). DOT/FAA/AR-97/61. Washington, D.C.: Federal Aviation Administration.

DOT (U.S. Department of Transportation). 1998. Aviation Security-Federal Aviation Administration. Washington, D.C.: Office of the Inspector General, U.S. Department of Transportation. Available on the Internet: http://www.dot.gov/oig/audits/av1998134.html

Ebco Technologies. 1998. TR9D Cyclotron System. Richmond, British Columbia: Ebco Technologies, Inc.

FAA (Federal Aviation Administration). 1997a. CTX-5000 and X-ray Blind Testing at San Francisco International Airport. Washington, D.C.: Federal Aviation Administration.

FAA. 1997b. Interim Report on the CTX-5000 SP Airport Demonstration. Washington, D.C.: Federal Aviation Administration.

FAA. 1997c. Final Report on CTX-5000 SP Operational Testing at Atlanta. Washington, D.C.: Federal Aviation Administration.

FAA. 1996a. Certification Test and Evaluation Report for the InVision CTX-5000 SP Explosives-Detection System. DOT/FAA/AR-96/86 (classified document). Washington, D.C.: Federal Aviation Administration.

FAA. 1996b. The Composition and Handling of Passenger Aircraft Cargo. DOT/FAA/AR-96/83. Washington, D.C.: Federal Aviation Administration.

FAA. 1995. Technical Center baggage inspection system: airport operational demonstration project. Washington, D.C.: Federal Aviation Administration.

FAA. 1992. Proposed Criteria for Certification of Explosives Detection Systems as Defined under 14 CFR 108.20. Washington, D.C.: Federal Aviation Administration.

Fainberg, A. 1998. Strategic Planning for Aviation Security. Presentation to the Panel on Assessment of Technologies Deployed to Improve Aviation Security, July 21, 1998, National Research Council, Washington, D.C.

Gibson, B.C., P.K. Van Staagen, and T.G. Miller. 1997. Blind Test Preliminary Report. October 30, 1997.

Gozani, T. 1988. Nuclear detection technology, from plutonium through coal to explosives. American Nuclear Society Transactions 56(3): 38.

Hussein, E.M.A. 1992. Detection of explosive materials using nuclear radiation: a critical review. Pp. 130–137 in X-Ray Detector Physics and Applications, R.B. Hoover, ed. Proceedings of the International Society for Optical Engineering, vol. 1736. Bellingham, Wash.: The International Society for Optical Engineering.

Jones, C.G. 1990. Environmental assessment of the thermal neutron activation explosive detection system for concourse use at U.S. airports. NUREG-1396. Washington, D.C.: U.S. Nuclear Regulatory Commission.

Lefevre, H.W. 1998. University of Oregon Archive: Blind tests of explosives detection in airline luggage; data analysis and results; candle analysis and other supporting software. Personal communication on CDROM (on file at the National Materials Advisory Board, August, 1998, National Research Council).

Lefevre, H.W., and J.C. Overley. 1998. Detection of Explosives through Fast Neutron Time-of-Flight Attenuation Measurements. Final Report for FAA Grant #94-G-020. Washington, D.C.: Federal Aviation Administration.

Micklich, B.J., C.L. Fink, L. Sagalovsky, F.L. Smith, and T.J. Yule. 1996. Feasibility of Explosives Detection Using Fast-Neutron Transmission Spectroscopy. Report No. DOT/FAA/CT-96/XX. Atlantic City International Airport, N.J.: FAA Technical Center.

Miller, T.G., and W.H. Makky. 1993. Application of Fast Neutron Spectroscopy/Radiography (FNS/R) to Airport Security. Pp. 184–196 in Neutrons, X-Rays, and Gamma Rays, J.M. Carpenter, D.B. Cline, R. Lanza, and D.F. Mildner, eds. Proceedings of the International Society for Optical Engineering, vol. 1737. Bellingham, Wash.: The International Society for Optical Engineering.

NRC (National Research Council). 1998. Configuration Management and Performance Verification of Explosives-Detection Systems. National Materials Advisory Board. Washington, D.C.: National Academy Press. Available on the Internet: http://www.nap.edu/readingroom/enter2.cgi?0309061962.html

NRC. 1997. Second Interim Report of the Committee on Commercial Aviation Security. National Materials Advisory Board. Washington, D.C.: National Academy Press. Available on the Internet: http://www2.nas.edu/nmab/262e.html

NRC. 1996. First Interim Report of the Committee on Commercial Aviation Security. National Materials Advisory Board. Washington, D.C.: National Academy Press.

NRC. 1993. Detection of Explosives for Commercial Aviation Security. National Materials Advisory Board. Washington, D.C.: National Academy Press.

NRC. In progress. Report of the Panel on Assessment of Technologies Deployed to Improve Aviation Security. National Materials Advisory Board, National Research Council, Washington, D.C.

Overley, J.C. 1998. Informal presentation to the Panel on Assessment of the Practicality of Pulsed Fast Neutron Transmission Spectroscopy for Aviation Security, Washington, D.C. June 27, 1998.

Overley, J.C. 1987. Element-sensitive computed tomography with fast neutrons. Nuclear Instruments and Methods in Physics Research B24/25: 1058–1062.

Overley, J.C. 1985. Determination of H, C, N, O content of bulk materials from neutron attenuation measurements. International Journal of Applied Radiation and Isotopes 36: 185–192.

Overley, J.C., M.S. Chmelik, R.J. Rasmussen, G.E. Sieger, R.M.S. Schofield, and H.W. Lefevre. 1997. Results of Blind Tests for Explosives in Luggage Using Fast-Neutron Transmission Spectroscopy. Pp. 305–310 in Proceedings of the Second Explosives Detection Technology Symposium and Aviation Security Technology Conference, W.H. Makky, ed. Washington, D.C.: Federal Aviation Administration.

Tensor Technology, Inc. 1998a. Advanced Studies on the Multi-Dimensional Neutron Radiometer (MDNR) Airline Security System. Washington, D.C.: Federal Aviation Administration.

Tensor Technology, Inc. 1998b. Neutron Transmission Studies for Airline Security. Final Report for FAA Grant 94-G-033. Washington, D.C.: Federal Aviation Administration.

Van Staagan, P.K., and T.G. Miller. 1998. Personal communication. November 16, 1998.

White House Commission on Aviation Safety and Security. 1996. Initial report to the President. September 9, 1996. Available on the Internet: http://www.aviationcommission.dot.gov/

White House Commission on Aviation Safety and Security. 1997. Final report to the President. February 12, 1997. Available on the Internet: http://www.aviationcommission.dot.gov/

# Biographical Sketches of Panel Members

**Patrick Griffin (chair)** is a principle member of the technical staff at Sandia National Laboratories and a member of the Committee on Commercial Aviation Security. At Sandia National Laboratories, he performs research in the areas of radiation modeling and simulation, neutron effects testing, radiation dosimetry, and radiation damage to materials. He is active in the standardization community and is the current chair of the American Society of Testing and Materials (ASTM) Subcommittee E10.05 on Nuclear Radiation Metrology.

**Robert Berkebile** is a consultant on air carrier operations with 41 years of experience, including the coordination of cargo and passenger baggage. Before his retirement, Mr. Berkebile was manager of customer services automation for USAirways and director of station services for US-Africa Airways. He is a member of the National Research Council (NRC) Panel on the Assessment of Technologies Deployed for Aviation Security, a panel under the Committee on Commercial Aviation Security.

**Homer Boynton** has extensive experience in security matters, including 25 years with the Federal Bureau of Investigation and 12 years with American Airlines as managing director of corporate security. He has chaired many advisory panels on airline security, including the Security Advisory Committee of the Air Transport Association and the Security Committee of the International Air Transport Association. He was a member of the Federal Aviation Administration (FAA) Research, Engineering and Development Advisory Committee.

**Len Limmer** retired in January 1998, closing a distinguished 26-year career with the Dallas/Fort Worth International Airport Board, a local government agency responsible for the administration, management, and operation of the world's second busiest airport. He has cross-functional experience in public safety, including security police, counterterrorism, explosives detection/disposal, crash rescue, structural fire, and hazardous materials remediation. He has been an officer of numerous Texas and national associations concerned with public safety in large metropolitan areas. He is a member of the NRC Panel on the Assessment of Technologies Deployed for Aviation Security, a panel under the Committee on Commercial Aviation Security.

**Harry Martz** is the leader of the nondestructive evaluation research and development thrust area for Lawrence Livermore National Laboratory (LLNL). He received a B.S. in 1979 from Siena College and an M.S. and Ph.D. in 1986 from Florida State University. For six years, he led the computed-tomography project at LLNL, applying computed-tomography and x-ray and proton radiography to material characterization and gamma-ray gauge techniques for treaty verification activities. His current projects include the use of nonintrusive x-ray and gamma-ray computed-tomography techniques as three-dimensional imaging tools to understand material properties and analyze radioactive waste forms. He has applied these techniques to the inspection of automobile and aircraft parts, reactor fuel tubes, high explosives, shape charges, and the contents of waste drums. The research and development in his group includes the design and construction of scanners and preprocessing, image reconstruction, and analysis algorithms. Dr. Martz chaired the NRC Panel on Configuration Management and Performance Verification of Explosives-Detection Systems, was a member of the Panel on Airport Passenger Screening, and is a member of the Committee on Commercial Aviation Security and the Panel on the Assessment of Technologies Deployed for Aviation Security.

**Clinton Oster, Jr.,** is a professor in the School of Public and Environmental Affairs at Indiana University. He has investigated the sequential steps, from basic research to applied research to development to commercial implementation to the progressively difficult decisions of continuing or terminating a project based on, among other things, eventual investment opportunities. Professor Oster's current research centers on aviation safety, transportation economics, international aviation, airport and airway infrastructure, environmental and natural resource policy, and environmental remediation. His most recent book, *Why Airplanes Crash: Aviation Safety in a Changing World*, was published in 1992. Professor Oster served on a Transportation Research Board study committee on highway speed limits and recently chaired a National Academy of Sciences study committee on civilian aviation employment. He has been a consultant on aviation and other transportation issues to the U.S. Department of Transportation, the FAA, the National Aeronautics and Space Administration, the European Bank for Reconstruction and Development, state and local governments, and private-sector companies in the United States, Canada, the United Kingdom, Russia, and Australia.